PCMP

Preparation for College Mathematics Program

G. Viglino

Ramapo College of New Jersey
April 2018

CONTENTS

Preface
PCMP
(Preparation for College Mathematics Program)

DESCRIPTION: The program is designed to prepare students for our Math General Education Courses. Successful completion of the program will enable one to register for MATH 101 (Math with Applications), MATH 104 (Mathematics for the Modern World), MATH 106 (Introduction to Math Modeling), or MATH 108 (Elementary Probability and Statistics).

TOPICS: Rational Numbers, Integer Exponents, Algebraic Expressions, Linear Equations and Inequalities, Quadratic Equations (factoring method), Lines, Systems of Equations, Functions, Percents, Radicals, Scientific Notation.

SUPPORT: Free tutoring is available at the Math/Science Tutoring Center in G-121.

The Program

To successfully complete PCMP a student has to pass six tests and a comprehensive final exam. Prior to attempting tests 2 through 6, a student must have passed its preceding test. One must pass test 6 prior to attempting the final exam.

Each of the six tests consists of 9 questions, at least seven of which must be answered correctly in order to pass the test.

- The final consists of 18 questions, at least fifteen of which must be answered correctly in order to pass the exam, and thereby complete the program.

- Sample tests along with visual presentations of its questions can be accessed at

  http://www.ramapo.edu/tas/faculty/giovanni-viglino/

  The problems on each test are similar to those on the sample test. Consequently:

> IF ONE UNDERSTANDS THE MATERIAL ON THE SAMPLE TESTS, THEN ONE WILL BE ABLE TO PASS EACH TEST, AND THE FINAL EXAM.

- Exams will be administered at the testing center. No limit is imposed on the number of times one can attempt to pass any of the exams.

Students have the option of "**unofficially**" taking tests anywhere at any time, with real-time feedback provided. While such exams **do not "count"**, they provide a valuable instructional component to the course.

I must express my deepest gratitude to Professor Scott Frees for developing the computer component of PCMP, and to Jefferson Sampson and Michael Revello for their continuing support of that component.

I also need to thank my math colleagues, Professors Marion Berger, Katarzyna Kowal, Sara Kuplinsky, and Ben Wang for their invaluable input throughout the development of the this module.

Sample Test 1
RATIONAL NUMBERS
(FRACTIONS)

Supplement for Sample Test 1 starts on page 11.

Question 1.1

Simplify:

$$\left(\frac{2}{5}\right)\left(\frac{3}{4}\right)$$

The correct answer is $\frac{3}{10}$. If you got it, move on to Question 1.2. If not, consider the following example:

EXAMPLE 1.1   Simplify:

$$\left(\frac{3}{8}\right)\left(\frac{5}{18}\right)$$

THE CANCELLATION PROPERTY:

$$\frac{a\cancel{c}}{b\cancel{c}} = \frac{a}{b}$$

(providing $c \neq 0$)

For example:

$$\frac{\cancel{3}\cdot 5}{8\cdot \cancel{3}\cdot 6} = \frac{5}{48}$$

Definition:

$$\frac{a}{b}\cdot\frac{c}{d} = \frac{ac}{bd}$$

For example:

$$\frac{1}{8}\cdot\frac{5}{6} = \frac{1\cdot 5}{8\cdot 6} = \frac{5}{48}$$

SOLUTION:

| Factor the 18: $18 = 3 \cdot 6$ |

$$\left(\frac{3}{8}\right)\left(\frac{5}{18}\right) = \frac{3}{8}\cdot\frac{5}{3\cdot 6}$$

$$= \frac{\cancel{3}}{8}\cdot\frac{5}{\cancel{3}\cdot 6}$$

| Cancel -- see margin |

$$= \frac{1}{8}\cdot\frac{5}{6}$$

| see definition in margin: | $= \frac{5}{48}$

Can you now manage Question 1.1:

Simplify: $\left(\frac{2}{5}\right)\left(\frac{3}{4}\right)$ Answer: $\frac{3}{10}$

If so, go to Question 1.2. If not:

1.1 Simplify   *Click-Video*

(a) $\left(\frac{7}{9}\right)\left(\frac{3}{14}\right)$ (b) $\frac{64}{30}\cdot\frac{15}{12}$

If you still can't solve Question 1.1: **Go to the tutoring center**.

Question 1.2

Simplify:

$$\frac{-3}{10}\cdot\frac{15}{21}\cdot\frac{7}{-5}$$

The correct answer is $\frac{3}{10}$. If you got it, move on to Question 1.3. If not, consider the following example:

$$-\frac{a}{b} = \frac{-a}{b} = \frac{a}{-b}$$

For example:

$$-\frac{6}{1} = \frac{-6}{1} = \frac{6}{-1}$$

The product (or quotient) of an **odd** number of negative numbers is **negative.** For example:

$$\frac{(-6)(2)(-30)}{1(-14)(9)} = -\frac{6 \cdot 2 \cdot 30}{14 \cdot 9}$$

The product (or quotient) of an **even** number of negative numbers is **positive.** For example:

$$(-2)(3)(-4) = 24$$

EXAMPLE 1.2   Simplify:

$$-6 \cdot \frac{2}{-14} \cdot \frac{-30}{9}$$

SOLUTION:

$$-6 \cdot \frac{2}{-14} \cdot \frac{-30}{9} = \frac{-6}{1} \cdot \frac{2}{-14} \cdot \frac{-30}{9}$$

$$= \frac{(-6)(2)(-30)}{1(-14)(9)}$$

see margin:
$$= -\frac{6 \cdot 2 \cdot 30}{14 \cdot 9} = -\frac{2 \cdot 3 \cdot 2 \cdot 3 \cdot 10}{2 \cdot 7 \cdot 3 \cdot 3} = -\frac{20}{7}$$

Can you now manage Question 1.2:

Simplify: $\dfrac{-3}{10} \cdot \dfrac{15}{21} \cdot \dfrac{7}{-5}$ Answer: $\dfrac{3}{10}$

If so, go to Question 1.3. If not:

1.2 Simplify   *Click-Video*

(a) $-\left(-\dfrac{4}{15}\right)\left(\dfrac{-5}{16}\right)$ (b) $\dfrac{-12}{25} \cdot \dfrac{-5}{-8}\left(-\dfrac{2}{3}\right)$

If you still can't solve Question 1.2: **Go to the tutoring center.**

Question 1.3   Simplify:

Note: The quotient

$$\frac{3}{8} \div \frac{3}{4}$$

can also be expressed in the form:

$$\frac{\frac{3}{8}}{\frac{3}{4}} \quad \text{or} \quad \frac{3/8}{3/4}$$

$$\frac{3}{8} \div \frac{3}{4}$$

The correct answer is $\dfrac{1}{2}$. If you got it, move on to Question 1.4. If not, consider the following example:

EXAMPLE 1.3   Simplify:

$$\left(\frac{15}{16} \div \frac{10}{8}\right) \quad \left[\text{or:} \quad \frac{\frac{15}{16}}{\frac{10}{8}} \quad \text{or:} \quad \frac{15/16}{10/8}\right]$$

SOLUTION:

$$\frac{\frac{15}{16}}{\frac{10}{8}} = \frac{15}{16} \cdot \frac{8}{10} = \frac{3 \cdot 5 \cdot 8}{2 \cdot 8 \cdot 5 \cdot 2} = \frac{3}{2 \cdot 2} = \frac{3}{4}$$

see margin

Definition:

$$\frac{a}{b} \div \frac{c}{d} = \frac{a}{b} \cdot \frac{d}{c}$$

In words:
   To divide:
   invert and multiply
For example:

invert

$$\frac{15}{16} \div \frac{10}{8} = \frac{15}{16} \cdot \frac{8}{10}$$

and multiply

Can you now manage Question 1.3:

Simplify: $\dfrac{3}{8} \div \dfrac{3}{4}$ Answer: $\dfrac{1}{2}$

If so, go to Question 1.4 below. If not:

1.3 Simplify   *Click-Video*

(a) $\dfrac{-14}{21} \div \dfrac{2}{9}$ (b) $\dfrac{\frac{8}{27}}{\frac{-16}{15}}$

If you still can't solve Question 1.3: **Go to the tutoring center**.

$\boxed{\text{Question 1.4}}$ ## Simplify:

$$\dfrac{\frac{3}{8}}{6}$$

The correct answer is $\dfrac{1}{16}$. If you got it, move on to Question 1.5. If not, consider the following example:

The expression
$$\dfrac{\frac{9}{2}}{15}$$
can also be expressed in the form
$$(9 \div 2) \div 15$$

EXAMPLE 1.4   Simplify:

$$\dfrac{\frac{9}{2}}{15}$$

SOLUTION:

$$\dfrac{\frac{9}{2}}{15} = \dfrac{\frac{9}{2}}{\frac{15}{1}} = \dfrac{9}{2} \cdot \dfrac{1}{15} = \dfrac{\cancel{3} \cdot 3}{2 \cdot \cancel{3} \cdot 5} = \dfrac{3}{10}$$

Can you now manage Question 1.4:

Simplify: $\dfrac{\frac{3}{8}}{6}$ Answer: $\frac{1}{16}$

If so, go to Question 1.5 below. If not:

1.4 Simplify   *Click-Video*

(a) $\dfrac{\frac{-14}{8}}{16}$ (b) $\dfrac{\frac{-10}{4}}{-15}$

If you still can't solve Question 1.4: **Go to the tutoring center**.

$\boxed{\text{Question 1.5}}$ ## Simplify:

$$\dfrac{-4}{\frac{-6}{7}}$$

The correct answer is $\dfrac{14}{3}$. If you got it, move on to Question 1.6. If not, consider the following example:

EXAMPLE 1.5   Simplify:

$$\frac{-15}{\frac{3}{-4}}$$

SOLUTION:

> the product of two negative numbers is positive

$$\frac{-15}{\frac{3}{-4}} = \frac{\frac{-15}{1}}{\frac{3}{-4}} = \frac{-15}{1} \cdot \frac{-4}{3} = \frac{3 \cdot 5 \cdot 4}{3} = 20$$

Can you now manage Question 1.5:

Simplify: $\dfrac{-4}{\frac{-6}{7}}$ Answer: $\dfrac{14}{3}$

If so, go to Question 1.6 below. If not:

1.5 Simplify   *Click-Video*

(a) $-\dfrac{14}{\frac{7}{3}}$ (b) $\dfrac{1}{\frac{8}{5}}$

If you still can't solve Question 1.5: **Go to the tutoring center**.

Question 1.6   **Simplify:**

$$\frac{\frac{2}{3} \cdot \frac{5}{4}}{\frac{-8}{9}}$$

The correct answer is $-\dfrac{15}{16}$. If you got it, move on to Question 1.7. If not, consider the following example:

EXAMPLE 1.6   Simplify:

$$\frac{\frac{-4}{7} \cdot \frac{21}{16}}{\frac{9}{8}}$$

SOLUTION:

$$\frac{\frac{-4}{7} \cdot \frac{21}{16}}{\frac{9}{8}} = \frac{-4}{7} \cdot \frac{21}{16} \cdot \frac{8}{9} = \frac{-4 \cdot 3 \cdot 7 \cdot 4 \cdot 2}{7 \cdot 4 \cdot 4 \cdot 3 \cdot 3} = -\frac{2}{3}$$

> invert and multiply

Can you now manage Question 1.6:

$$\text{Simplify:} \quad \dfrac{\dfrac{2}{3}\cdot\dfrac{5}{4}}{\dfrac{-8}{9}} \qquad\qquad \text{Answer:} -\dfrac{15}{16}$$

If so, go to Question 1.7 below. If not:

1.6 Simplify   *Click-Video*

(a) $\dfrac{\dfrac{2}{5}\cdot\dfrac{7}{3}}{-\dfrac{28}{10}}$ (b) $\dfrac{\dfrac{5}{12}}{\dfrac{1}{2}\cdot\dfrac{15}{3}}$

If you still can't solve Question 1.6: **Go to the tutoring center**.

| **Question 1.7** | **Simplify:** |

$$\dfrac{2}{3}+\dfrac{7}{3}-\dfrac{4}{3}$$

The correct answer is $\dfrac{5}{3}$. If you got it, move on to Question 1.8. If not, consider the following example:

EXAMPLE 1.7   Simplify:

$$-\dfrac{3}{5}+\dfrac{4}{5}-\dfrac{2}{5}$$

SOLUTION:

To add (or subtract) fractions with the **same denominator**, add (or subtract) the numerators of those fractions and place the result over that denominator. In other words:

$$\dfrac{a}{d}+\dfrac{b}{d}=\dfrac{a+b}{d}$$

and $\dfrac{a}{d}-\dfrac{b}{d}=\dfrac{a-b}{d}$

$$-\dfrac{3}{5}+\dfrac{4}{5}-\dfrac{2}{5}=\dfrac{-3+4-2}{5}=\dfrac{-1}{5}=-\dfrac{1}{5}$$

(see margin)

Can you now manage Question 1.7:

$$\text{Simplify:} \ \dfrac{2}{3}+\dfrac{7}{3}-\dfrac{4}{3} \qquad\qquad \text{Answer:} \ \tfrac{5}{3}$$

If so, go to Question 1.8 below. If not:

1.7 Simplify   *Click-Video*

(a) $\dfrac{3}{7}+\dfrac{5}{7}+\dfrac{-5}{7}$ (b) $-\dfrac{2}{9}-\dfrac{-5}{9}+\dfrac{3}{9}$

If you still can't solve Question 1.7: **Go to the tutoring center**.

Question 1.8

Simplify:

$$\frac{5}{6} + \frac{2}{7}$$

The correct answer is $\frac{47}{42}$. If you got it, move on to Question 1.9. If not, consider the following example:

EXAMPLE 1.8   Simplify:

$$\frac{2}{3} - \frac{5}{4}$$

SOLUTION: The least common denominator of $\frac{2}{3}$ and $\frac{5}{4}$ is 12 (see

The **least common denominator** of two or more fractions is the smallest integer that is divisible by the denominator of each fraction. For example, 12 is the least common denominator of $\frac{2}{3}$ and $\frac{5}{4}$ (12 is the smallest integer divisible by both 3 and 4).

margin). So:

$$\frac{2}{3} - \frac{5}{4} = \frac{2 \cdot 4}{3 \cdot 4} - \frac{5 \cdot 3}{4 \cdot 3} = \frac{8}{12} - \frac{15}{12} = \frac{8 - 15}{12} = -\frac{7}{12}$$

see margin

common denominator

Can you now manage Question 1.8:

Simplify: $\frac{5}{6} + \frac{2}{7}$ Answer: $\frac{47}{42}$

If so, go to Question 1.9 below. If not:

1.8 Simplify   *Click-Video*

(a) $\frac{3}{7} - \frac{5}{14} + \frac{1}{2}$ (b) $-\frac{2}{9} - \frac{-2}{3} + \frac{5}{6}$

If you still can't solve Question 1.8: **Go to the tutoring center**.

Question 1.9

Simplify:

$$\frac{\frac{9}{2}}{9} - \frac{4}{-\frac{1}{2}}$$

The correct answer is $\frac{17}{2}$. If you got it, great. If not, consider the following example:

EXAMPLE 1.9   Simplify:

$$\frac{\frac{-7}{3}}{1+\frac{1}{3}}+\frac{2-\frac{1}{3}}{\frac{2}{3}}$$

Looks intimidating? Just take it one step at a time.

SOLUTION: One approach:

$$\frac{\frac{-7}{3}}{1+\frac{1}{3}}+\frac{2-\frac{1}{3}}{\frac{2}{3}}=\frac{3\left(\frac{-7}{3}\right)}{3\left(1+\frac{1}{3}\right)}+\frac{3\left(2-\frac{1}{3}\right)}{3\left(\frac{2}{3}\right)}=\frac{-7}{3+1}+\frac{6-1}{2}$$

There are many different approaches one can take to solve this problem. We offer two for your consideration

$$=\frac{-7}{4}+\frac{5}{2}=\frac{-7}{4}+\frac{10}{4}=\frac{3}{4}$$

Another approach:

$$\frac{\frac{-7}{3}}{1+\frac{1}{3}}+\frac{2-\frac{1}{3}}{\frac{2}{3}}=\frac{\frac{-7}{3}}{\frac{3}{3}+\frac{1}{3}}+\frac{\frac{6}{3}-\frac{1}{3}}{\frac{2}{3}}=\frac{\frac{-7}{3}}{\frac{4}{3}}+\frac{\frac{5}{3}}{\frac{2}{3}}=\frac{-7}{3}\cdot\frac{3}{4}+\frac{5}{3}\cdot\frac{3}{2}$$

$$=\frac{-7}{4}+\frac{5}{2}$$

$$=\frac{-7}{4}+\frac{10}{4}$$

$$=\frac{-7+10}{4}=\frac{3}{4}$$

Can you now manage question 1.9:

Simplify: $\dfrac{\frac{9}{2}}{9}-\dfrac{4}{-\frac{1}{2}}$ Answer: $\frac{17}{2}$

If not:

1.9 Simplify   **Click-Video**

(a) $\dfrac{3}{4}+\dfrac{\frac{1}{2}}{\frac{1}{3}\cdot 6}-\dfrac{1}{6}$ (b) $1+\dfrac{1}{1+\frac{1}{2}}-\dfrac{1+\frac{1}{2}}{2}$

If you still can't solve Question 1.9: **Go to the tutoring center**.

	SUMMARY	

MULTIPLYING FRACTIONS	**DEFINITION:** $$\frac{a}{b} \cdot \frac{c}{d} = \frac{ac}{bd}$$
DIVIDING FRACTIONS	**DEFINITION:** $$\frac{\dfrac{a}{b}}{\dfrac{c}{d}} = \frac{a}{b} \cdot \frac{d}{c}$$ **(invert and multiply)**
ADDING AND SUBTRACTING FRACTIONS WITH THE SAME DENOMINATOR	**DEFINITION:** $$\frac{a}{d} + \frac{b}{d} = \frac{a+b}{d} \qquad \text{and} \qquad \frac{a}{d} - \frac{b}{d} = \frac{a-b}{d}$$
LEAST COMMON DENOMINATOR OF FRACTIONS	The smallest positive integer that is divisible by the denominator of each fraction.
ADDING OR SUBTRACTING FRACTIONS WITH DIFFERENT DENOMINATORS	Determine the least common denominator of the fractions. For each fraction, multiply its numerator and denominator by the same number so that its denominator becomes the least common denominator. Then add or subtract the resulting fractions.
FRACTIONS AND MINUS SIGNS	$$\frac{-a}{b} = \frac{a}{-b} = -\frac{a}{b}$$

CANCELLATION PROPERTY	$$\frac{a\cancel{c}}{b\cancel{c}} = \frac{a}{b} \quad \text{If } c \neq 0$$
	You can **ONLY** cancel a **common FACTOR.** For example:
NOTE:	$$\frac{\cancel{7} \cdot 3}{\cancel{7} \cdot 5} = \frac{3}{5}$$ Cancel the 7, which is a factor of both the numerator and the denominator.

and:

$$\frac{26}{16} = \frac{2 \cdot 13}{2 \cdot 8} = \frac{13}{8}$$

You **CANNOT** cancel the 2 in $\frac{2 + 13}{2 \cdot 8}$ **!!!!!**

Yes, the 2 is a factor in the denominator (2 **times** something) but it is not a factor in the numerator (the numerator is **not** 2 times something -- it is 2 plus something).

On the other hand:

$$\frac{2 + 2a}{2 + 4b} = \frac{\cancel{2}(1 + a)}{\cancel{2}(1 + 2b)} = \frac{1 + a}{1 + 2b}$$

a common factor

It bears repeating:

You can **ONLY** cancel a **common FACTOR:**

$$\frac{a\cancel{c}}{b\cancel{c}} = \frac{a}{b}$$

(providing c is not zero)

ADDITIONAL PROBLEMS

1.1 $\left(\frac{3}{5}\right)\left(\frac{10}{6}\right)$ Answer: 1	1.2 $\frac{-3}{5} \cdot \frac{15}{-2}$ Answer: $\frac{9}{2}$	2.1 $2\left(\frac{8}{6}\right)\left(\frac{3}{12}\right)$ Answer: $\frac{2}{3}$	2.2 $-\frac{3}{2} \cdot \frac{4}{9} \cdot \frac{1}{-2}$ Answer: $-\frac{1}{3}$
3.1 $\frac{2}{5} \div \frac{10}{6}$ Answer: $\frac{6}{25}$	3.2 $\dfrac{\frac{1}{2}}{\frac{3}{-2}}$ Answer: $-\frac{1}{3}$	4.1 $\dfrac{\frac{3}{5}}{6}$ Answer: $\frac{1}{10}$	4.2 $\dfrac{\frac{-5}{6}}{-4}$ Answer: $\frac{5}{24}$

5.1 $\dfrac{3}{\frac{6}{5}}$ Answer: $\frac{5}{2}$	5.2 $-\dfrac{2}{\frac{4}{3}}$ Answer: $\frac{3}{2}$	6.1 $\dfrac{\frac{4}{5}\cdot\frac{15}{2}}{\frac{3}{5}}$ Answer: 10	6.2 $\dfrac{-3\cdot\frac{2}{6}}{-\frac{5}{3}}$ Answer: $\frac{3}{5}$
6.3 $\dfrac{\frac{3}{2}\left(-\frac{2}{5}\right)}{-6}$ Answer: $\frac{1}{10}$	6.4 $\dfrac{-\frac{2}{3}\cdot\frac{5}{6}\cdot\frac{12}{4}}{-\frac{9}{5}\cdot\frac{1}{2}}$ Answer: $\frac{50}{27}$	6.5 $\dfrac{\frac{\frac{2}{3}\cdot\frac{1}{5}}{2}}{2\cdot\frac{1}{3}}$ Answer: $\frac{1}{10}$	6.6 $\dfrac{3+\frac{1}{2}}{\frac{\frac{1}{5}}{2}}$ Answer: 35
7.1 $\dfrac{1}{5}-\dfrac{3}{5}+\dfrac{4}{5}$ Answer: $\frac{2}{5}$	7.2 $-\dfrac{-2}{7}+\dfrac{5}{7}-\dfrac{3}{7}$ Answer: $\frac{4}{7}$	8.1 $\dfrac{3}{10}+\dfrac{2}{5}$ Answer: $\frac{7}{10}$	8.2 $\dfrac{1}{6}-\dfrac{2}{3}+\dfrac{3}{18}$ Answer: $-\frac{1}{3}$
9.1 $\dfrac{\frac{1}{5}-\frac{2}{5}}{\frac{2}{3}+\frac{1}{3}}$ Answer: $-\frac{1}{5}$	9.2 $\left(\dfrac{\frac{1}{3}}{\frac{2}{3}+\frac{4}{3}}\right)\left(\dfrac{-3}{\frac{1}{2}}\right)$ Answer: -1	9.3 $\dfrac{\frac{3}{10}+\frac{2}{5}}{-\frac{1}{10}}-\dfrac{3}{-2}$ Answer: $-\frac{11}{2}$	9.4 $\dfrac{\left(\dfrac{\frac{1}{6}-\frac{2}{3}+\frac{1}{4}}{-\frac{3}{2}}\right)}{\frac{2}{3}}$ Answer: $\frac{1}{4}$

Sample Test 1 Supplement

RATIONAL NUMBERS

A **rational number** (fraction) is an expression of the form

$$\frac{a}{b} \quad \text{or} \quad a/b$$

where a is any integer, and b is any **nonzero** integer.

*a is said to be the **numerator** of the fraction $\frac{a}{b}$, and b its **denominator**.*

DEFINITION 1.1

EQUALITY OF RATIONAL NUMBERS

The rational numbers $\frac{a}{b}$ and $\frac{c}{d}$ are said to be equal, and we write $\frac{a}{b} = \frac{c}{d}$, if $ad = bc$.

In other words:

$\frac{a}{b} = \frac{c}{d}$ if the "**cross products**" are equal: $\frac{a}{b} \diagdown\!\!\!\diagup \frac{c}{d}$

For example:

$$\frac{3}{5} = \frac{6}{10} \quad \text{since} \quad 3 \cdot 10 = 5 \cdot 6$$

$$\frac{-9}{3} = \frac{12}{-4} \quad \text{since} \quad (-9)(-4) = 3 \cdot 12$$

$$\frac{4}{2} = 2 \quad \text{since} \quad 4 \cdot 1 = 2 \cdot 2$$

$$\frac{4}{2} = \frac{2}{1}$$

and $\quad \dfrac{2}{a} = \dfrac{4}{2a} \quad \text{since} \quad 2(2a) = 4a$

When we write $\dfrac{2}{a} = \dfrac{4}{2a}$, it is understood that $a \neq 0$.

While we're at it, why can't we "divide" by zero? Consider "$\dfrac{12}{0}$". If it were to equal a number, say "$\dfrac{12}{0} = a$," then it would have to follow that $12 = a(0)$, which cannot be.

Of course, in the above argument the number 12 can be replaced by any nonzero number, so we see that "$\dfrac{b}{0}$" is **meaningless** for any $b \neq 0$.

*Note that $\frac{0}{12} = 0$ since $0 = 12(0)$. It is **only** the denominator that cannot be zero.*

"$\dfrac{0}{0}$" is also **meaningless**; for to say that "$\dfrac{0}{0} = a$" is to say that

$0 = a(0)$, and since $a(0) = 0$ for *any a,* we could associate any number we please with "$\frac{0}{0}$", and that we cannot permit.

The following result follows directly from Definition 1.1:

THEOREM 1.1   For any rational number $\frac{a}{b}$ and any $c \neq 0$:

$$\frac{ac}{bc} = \frac{a}{b}$$

PROOF: $\frac{ac}{bc} = \frac{a}{b}$ since $acb = bca$

Equations are "two way streets." If you read the equation in Theorem 1.1 from left to right, you get the **cancellation property**:

CANCELLATION PROPERTY

For any a, and for any b and c distinct from zero:

$$\frac{a\cancel{c}}{b\cancel{c}} = \frac{a}{b}$$

∟ _ *cancel*

Reading the same equation from right to left, you get the **build-up property:**

BUILD-UP PROPERTY

For any a, and for any b and c distinct from zero:

$$\frac{a}{b} = \frac{ac}{bc}$$

WARNING: The cancellation property has nothing to do with sums. In particular, there can be **NO** cancellation in any of the following three expressions:

$$\frac{5+8}{5 \cdot 8} \qquad \frac{c}{c+5} \qquad \frac{2+a}{a}$$

Here is the **ONLY** time you can cancel (if $c \neq 0$):

$$\frac{\cancel{c}\,\textbf{times something}}{\cancel{c}\,\textbf{times something else}} = \frac{\textbf{something}}{\textbf{something else}}$$

In other words, c must be a **nonzero factor** of both the numerator and the denominator.

LOWEST TERMS

A fraction is said to be in **lowest terms** if the numerator and denominator have no common factors. For example, $\frac{3}{8}$ is in lowest terms, while $\frac{6}{8}$ is not, since 6 and 8 share a common factor: 2.

EXAMPLE 1.1 Reduce to lowest terms.

$$(a)\ \frac{6}{8} \qquad (b)\ \frac{6a}{8a^2} \qquad (c)\ \frac{-15(2+c)}{-5(2+c)}$$

You **CANNOT** do any canceling here: $\frac{2+3}{2+4}$

Or here: $\frac{3+2a}{8a^2}$

Or here: $\frac{-15(2+c)}{-5+(2+c)}$

SOLUTION:

(a) $\dfrac{6}{8} = \dfrac{\not{2} \cdot 3}{\not{2} \cdot 4} = \dfrac{3}{4}$

(b) $\dfrac{6a}{8a^2} = \dfrac{3(2a)}{4a(2a)} = \dfrac{3}{4a}$ (if $a \neq 0$)

(c) $\dfrac{-15(2+c)}{-5(2+c)} = \dfrac{-15}{-5} = \dfrac{-5(3)}{-5(1)} = \dfrac{3}{1} = 3$ [if $c \neq -2$ (why?)]

CHECK YOUR UNDERSTANDING 1.1

Reduce to lowest terms (assuming no denominator is zero).

$$(a)\ \frac{24}{30} \qquad (b)\ \frac{-92}{44} \qquad (c)\ \frac{15(2+c)}{10(c+2)} \qquad (d)\ \frac{7b(a+c)}{b(-a-c)}$$

Answers: (a) $\dfrac{4}{5}$ (b) $\dfrac{-23}{11}$ (c) $\dfrac{3}{2}$ (d) -7

Solution: Page 107.

To multiply two fractions, simply multiply the numerators and the denominators:

MULTIPLICATION

$$\frac{a}{b} \cdot \frac{c}{d} = \frac{ac}{bd}$$

Notice that the above definition "respects" integer multiplication (as it must, since the set of integers is contained in the set of rational numbers). For example, we have:

$$5 \cdot 7 = \frac{5}{1} \cdot \frac{7}{1} = \frac{5 \cdot 7}{1 \cdot 1} = \frac{35}{1} = 35$$

Notice also that for any rational number $\dfrac{a}{b}$:

$$1 \cdot \frac{a}{b} = \frac{1}{1} \cdot \frac{a}{b} = \frac{1 \cdot a}{1 \cdot b} = \frac{a}{b}$$

EXAMPLE 1.2 Multiply, and reduce to lowest terms.

$$(a)\ \frac{-7}{2} \cdot \frac{3}{5} \qquad (b)\ \frac{2}{6} \cdot \frac{9}{10} \qquad (c)\ \frac{a}{3} \cdot \frac{15}{-a^2}$$

SOLUTION:

(a) $\dfrac{-7}{2} \cdot \dfrac{3}{5} = \dfrac{-7 \cdot 3}{2 \cdot 5} = \dfrac{-21}{10}$

(b) One way: $\dfrac{2}{6} \cdot \dfrac{9}{10} = \dfrac{2 \cdot 9}{6 \cdot 10} = \dfrac{\cancel{2} \cdot \cancel{3} \cdot 3}{\cancel{2} \cdot \cancel{3} \cdot 2 \cdot 5} = \dfrac{3}{10}$

Quicker is better: $\dfrac{2}{\cancel{6}} \cdot \dfrac{\overset{3}{\cancel{9}}}{10} = \dfrac{1}{\cancel{3}} \cdot \dfrac{\cancel{9}}{10} = \dfrac{3}{10}$

(c) $\dfrac{a}{3} \cdot \dfrac{15}{-a^2} = \dfrac{3 \cdot 5a}{3(-1)a \cdot a} = \dfrac{5}{-a} = -\dfrac{5}{a}$, **OR:** $\dfrac{\cancel{a}}{\cancel{3}} \cdot \dfrac{\overset{5}{\cancel{15}}}{-a^{\cancel{2}}} = \dfrac{5}{-a} = -\dfrac{5}{a}$

In the margin:

$\dfrac{-a}{b} = \dfrac{a}{-b}$ since the cross products are equal: $(-a)(-b) = ab$.

$\dfrac{-a}{b} = -\dfrac{a}{b}$ is saying that $\dfrac{-a}{b}$ is the additive inverse of $\dfrac{a}{b}$ (as you probably already know: $\dfrac{a}{b} + \dfrac{-a}{b} = 0$.

In general (see margin), for any $b \neq 0$:

$$\dfrac{-a}{b} = \dfrac{a}{-b} = -\dfrac{a}{b}$$

To multiply three or more fractions, you still multiply numerators and multiply denominators. For example:

$$\dfrac{1}{2} \cdot \dfrac{-3}{5} \cdot \dfrac{2}{7} = \dfrac{\cancel{2}(-3)}{\cancel{2} \cdot 5 \cdot 7} = \dfrac{-3}{35} = -\dfrac{3}{35}$$

and: $\left(\dfrac{a}{b}\right)\left(\dfrac{3}{a}\right)(7b^2)\left(\dfrac{1}{3}\right) = \dfrac{\cancel{a} \cdot \cancel{3} \cdot 7 \cdot \cancel{b} \cdot b}{\cancel{a} \cdot \cancel{3} \cdot \cancel{b}} = 7b$

CHECK YOUR UNDERSTANDING 1.2

Multiply, and reduce to lowest terms.

(a) $\left(\dfrac{-5}{21}\right)\left(\dfrac{14}{10}\right)$

(b) $\dfrac{9}{5} \cdot \dfrac{15}{8} \cdot 12 \cdot \dfrac{2}{6}$

Answers: (a) $-\dfrac{1}{3}$ (b) $\dfrac{27}{2}$

Solution: Page 107.

DIVISION

To divide fractions, *invert* and *multiply*

$$\dfrac{\dfrac{a}{b}}{\dfrac{c}{d}} = \dfrac{a}{b} \cdot \dfrac{d}{c}$$

EXAMPLE 1.3   Divide, and reduce to lowest terms.

$$\text{(a) } \dfrac{\frac{2}{5}}{\frac{-7}{10}} \qquad\qquad \text{(b) } \dfrac{\frac{1}{2}}{4}$$

$$\text{(c) } \dfrac{8}{\frac{2}{3}} \qquad\qquad \text{(d) } \dfrac{\frac{a}{2b}}{\frac{4a}{b}}$$

SOLUTION:

(a) $\dfrac{\frac{2}{5}}{\frac{-7}{10}} = \dfrac{2}{5} \cdot \dfrac{10}{-7} = \dfrac{2 \cdot 10}{5(-7)} = \dfrac{2 \cdot 2 \cdot 5}{-7 \cdot 5} = \dfrac{4}{-7} = -\dfrac{4}{7}$

$\underset{\text{invert}}{\underbrace{\qquad}}$

(b) $\dfrac{\frac{1}{2}}{4} = \dfrac{\frac{1}{2}}{\frac{4}{1}} = \dfrac{1}{2} \cdot \dfrac{1}{4} = \dfrac{1}{8}$

$\underset{\text{invert}}{\underbrace{\qquad}}$

(c) $\dfrac{8}{\frac{2}{3}} = \dfrac{\frac{8}{1}}{\frac{2}{3}} = \dfrac{\overset{4}{8}}{1} \cdot \dfrac{3}{2} = 12$

(d) $\dfrac{\frac{a}{2b}}{\frac{4a}{b}} = \dfrac{a}{2b} \cdot \dfrac{b}{4a} = \dfrac{1}{8}$ $\left(\begin{array}{l}\text{with the unerstanding}\\ \text{that } a \neq 0 \text{ and } b \neq 0\end{array}\right)$

CHECK YOUR UNDERSTANDING 1.3

Divide, and reduce to lowest terms. [In (d), assume $a + b \neq 0$)].

$$\text{(a) } \dfrac{\frac{2}{3}}{\frac{8}{9}} \qquad \text{(b) } \dfrac{\frac{2}{3}}{9} \qquad \text{(c) } \dfrac{2}{\frac{8}{9}} \qquad \text{(d) } \dfrac{\frac{2a + 2b}{4}}{\frac{a + b}{8}}$$

Answers: (a) $\frac{3}{4}$ (b) $\frac{2}{27}$ (c) $\frac{9}{4}$ (d) 4

Solution: Page 107.

$$\boxed{\textbf{ADDING FRACTIONS}}$$

We start off by defining the addition of fractions that have a common denominator, and define their sum to be the sum of the numerators over the common denominator:

ADDITION
$$\frac{a}{d} + \frac{b}{d} = \frac{a+b}{d}$$

Note that with the above definition, we have:

$$\frac{5}{1} + \frac{7}{1} = \frac{5+7}{1} = 12 \text{ (as it should be!)}$$

Of course, addition may be extended to three or more fractions:

$$\frac{9}{5} + \frac{4}{5} + \frac{-3}{5} = \frac{9+4-3}{5} = \frac{10}{5} = 2$$

CHECK YOUR UNDERSTANDING 1.4

Sum, and reduce to lowest terms.

(a) $\dfrac{-3}{9} + \dfrac{2}{9} + \dfrac{19}{9}$ (b) $\dfrac{3}{a} + \dfrac{2}{a} + \dfrac{-5}{a}$ (c) $\dfrac{-3a}{4} + \dfrac{2(a-1)}{4} + \dfrac{5}{4}$

Answers: (a) 2   (b) 0 (assuming $a \neq 0$)   (c) $\dfrac{-a+3}{4}$

Solution: Page 107.

So far so good. But how does one go about adding fractions that do not share a common denominator? The answer should be apparent:

> Express each fraction in a form such that they share a common denominator, and then add.

For example, to determine the sum:

$$\frac{7}{2} + \frac{5}{3}$$

express both fractions with 6 as the denominator:

$$\frac{7}{2} = \frac{7 \cdot 3}{2 \cdot 3} = \frac{21}{6} \qquad \text{and} \qquad \frac{5}{3} = \frac{5 \cdot 2}{3 \cdot 2} = \frac{10}{6}$$

and then add:

$$\frac{7}{2} + \frac{5}{3} = \frac{21}{6} + \frac{10}{6} = \frac{31}{6}$$

Representing $\dfrac{7}{2}$ and $\dfrac{5}{3}$ as fractions with a common denominator distinct from 6 will not alter the sum. For example:

$$\frac{7}{2} + \frac{5}{3} = \frac{7 \cdot 6}{2 \cdot 6} + \frac{5 \cdot 4}{3 \cdot 4} = \frac{42}{12} + \frac{20}{12} = \frac{62}{12} = \frac{31 \cdot 2}{6 \cdot 2} = \frac{31}{6}$$

But it is best to represent the fractions with a common denominator as small as possible. That denominator is called the **least common denominator (LCD)**, and it is simply the least common multiple of the denominators of the fractions being summed.

LCD

EXAMPLE 1.4   Sum, and reduce to lowest terms.

$$\frac{5}{6} + \frac{3}{14} + \frac{1}{4}$$

SOLUTION: To determine the sum $\frac{5}{6} + \frac{3}{14} + \frac{1}{4}$, we first find the LCD of $\frac{5}{6} + \frac{3}{14} + \frac{1}{4}$ (equivalently, the least common multiple of the denominators 6, 14, and 4):

> You can also consider multiples of 14 and stop when you get to one that is divisible by both 4 and 6; namely: 84.

$$
\begin{array}{r}
6 = 2 \cdot 3 \\
14 = 2 \cdot 7 \\
4 = 2^2 \\
\hline
\text{LCD} = 2^2 \cdot 3 \cdot 7 = 84
\end{array}
$$

Looking at the above form you can easily see that:

To go from $6 = 2 \cdot 3$ to LCD $= 2^2 \cdot 3 \cdot 7$, you need one more 2 and a 7, which is to say, you have to multiply $6 = 2 \cdot 3$ by $2 \cdot 7 = 14$.

> Or: "6 goes into 84 fourteen times."

To go from $14 = 2 \cdot 7$ to LCD $= 2^2 \cdot 3 \cdot 7$, you need one more 2 and a 3, and therefore multiply $14 = 2 \cdot 7$ by $2 \cdot 3 = 6$.

To go from $4 = 2^2$ to LCD $= 2^2 \cdot 3 \cdot 7$, you have to multiply by $3 \cdot 7 = 21$.

Bringing us to:

$$\frac{5}{6} + \frac{3}{14} + \frac{1}{4} = \frac{5 \cdot 14}{6 \cdot 14} + \frac{3 \cdot 6}{14 \cdot 6} + \frac{1 \cdot 21}{4 \cdot 21}$$

$$= \frac{70}{84} + \frac{18}{84} + \frac{21}{84} = \frac{70 + 18 + 21}{84} = \frac{109}{84}$$

To see that the answer is in lowest terms, you need but observe that no prime in the decomposition of 84 (2, 3, and 7) is a divisor of 109.

CHECK YOUR UNDERSTANDING 1.5

Sum, and reduce to lowest terms.

(a) $\frac{-3}{18} + \frac{1}{12} + \frac{3}{8}$ (b) $\frac{-2}{15} + \frac{1}{5} + \frac{1}{3}$

Answers: (a) $\frac{7}{24}$ (b) $\frac{2}{5}$

Answer: Page 107.

EXAMPLE 1.5 Perform the indicated operations and reduce to lowest terms.

$$\text{(a) } \left(\frac{1}{3} - \frac{4}{9}\right)\left(2 + \frac{1}{4}\right) \qquad \text{(b) } \frac{1 + \frac{1}{3}}{3} \qquad \text{(c) } \frac{3 + \frac{2}{3}}{2 - \frac{1}{9}}$$

SOLUTION:

(a) $\left(\frac{1}{3} - \frac{4}{9}\right)\left(2 + \frac{1}{4}\right) = \left(\frac{3}{9} - \frac{4}{9}\right)\left(\frac{8}{4} + \frac{1}{4}\right)$

$$= \left(\frac{3-4}{9}\right)\left(\frac{8+1}{4}\right) = \frac{-1}{9} \cdot \frac{9}{4} = -\frac{1}{4}$$

A fraction such as

$$\frac{1 + \frac{1}{3}}{3}$$

whose numerator or denominator contains fractions is said to be a **compound fraction**.

(b) $\dfrac{1 + \dfrac{1}{3}}{3} = \dfrac{\dfrac{3}{3} + \dfrac{1}{3}}{3} = \dfrac{\dfrac{4}{3}}{\dfrac{3}{1}} = \dfrac{4}{3} \cdot \dfrac{1}{3} = \dfrac{4}{9}$

(c) One way: $\dfrac{3 + \dfrac{2}{3}}{2 - \dfrac{1}{9}} = \dfrac{\dfrac{9}{3} + \dfrac{2}{3}}{\dfrac{18}{9} - \dfrac{1}{9}} = \dfrac{\dfrac{11}{3}}{\dfrac{17}{9}} = \dfrac{11}{3} \cdot \dfrac{9}{17} = \dfrac{11 \cdot 3}{17} = \dfrac{33}{17}$

Another way is to multiply both the numerator and denominator by 9 so as to "un-compound" the given compound fraction:

$$\frac{3 + \frac{2}{3}}{2 - \frac{1}{9}} = \frac{9\left(3 + \frac{2}{3}\right)}{9\left(2 - \frac{1}{9}\right)} = \frac{9 \cdot 3 + 9 \cdot \frac{2}{3}}{9 \cdot 2 - 9 \cdot \frac{1}{9}} = \frac{27 + 6}{18 - 1} = \frac{33}{17}$$

CHECK YOUR UNDERSTANDING 1.6

Perform the indicated operations and reduce to lowest terms.

$$\text{(a) } \frac{\left(-2 + \frac{1}{3}\right)\left(1 - \frac{1}{2}\right)}{1 + \frac{1}{2}} \qquad\qquad \text{(b) } \frac{\frac{3b}{a+b} - \frac{a}{2a+2b}}{\frac{4}{a+b}}$$

Answers: (a) $-\dfrac{5}{9}$ (b) $\dfrac{6b-a}{8}$

Solution: Page 108.

Sample Test 2
INTEGER EXPONENTS

Supplement for Sample Test 2 starts on page 27.

Question 2.1

Simplify:

$$4^2 + 2^4$$

The correct answer is 32. If you got it, move on to Question 2.2. If not, consider the following example:

EXAMPLE 2.1   Simplify:

$$2^3 + 3^2$$

SOLUTION:

$$2^3 + 3^2 = 2 \cdot 2 \cdot 2 + 3 \cdot 3 = 8 + 9 = 17$$

see margin

Definition. For any number a and any positive integer n:

$$a^n = \overbrace{a \cdot a \cdots a}^{n\text{-times}}$$

For example:

$$2^3 = \overbrace{2 \cdot 2 \cdot 2}^{3\text{-times}}$$

Can you now manage Question 2.1:

Simplify: $4^2 + 2^4$ Answer: 32

If so, go to Question 2.2. If not:

2.1 Simplify   *Click-Video*
(a) $3^2 + 2^4$ (b) $2^2 - 2^3 + 3^2$

If you still can't solve Question 2.1: **Go to the tutoring center.**

Question 2.2

Simplify:

$$\frac{-5^2}{3} + (-4)^2$$

The correct answer is $\frac{23}{3}$. If you got it, move on to Question 2.3. If not, consider the following example:

EXAMPLE 2.2   Simplify:

$$\frac{3}{-2^2} + (-2)^2$$

SOLUTION:

$$\frac{3}{-2^2} + (-2)^2 = \frac{3}{-4} + 4 = -\frac{3}{4} + 4 = -\frac{3}{4} + \frac{16}{4} = \frac{13}{4}$$

see margin

Yes, a negative number raised to an even power is positive. For example:

$$(-2)^2 = 4$$

On the other hand -2^2 is **not** positive — it is minus the square of 2:

$$-2^2 = -(2^2) = -4$$

(only the 2 is being squared)

Can you now manage Question 2.2:

Simplify: $\frac{-5^2}{3} + (-4)^2$ Answer: $\frac{23}{3}$

If so, go to Question 2.3. If not:

2.2 Simplify   *Click-Video*

(a) $\dfrac{-2^2}{(-3)^2}$ (b) $\left(-\dfrac{1}{2}\right)^2 - 3^2$

If you still can't solve Question 2.2: **Go to the tutoring center.**

Question 2.3

Simplify:

$$\frac{(3+2)^2}{(2\cdot 5)^3}$$

The correct answer is $\dfrac{1}{40}$. If you got it, move on to Question 2.4. If not, consider the following example:

EXAMPLE 2.3   Simplify:

$$\frac{(2\cdot 3)^2}{(2+1)^3}$$

SOLUTION: One approach:

$$\frac{(2\cdot 3)^2}{(2+1)^3} = \frac{2\cdot 3 \cdot 2 \cdot 3}{3\cdot 3\cdot 3} = \frac{4}{3}$$

Another approach:

$$\frac{(2\cdot 3)^2}{(2+1)^3} = \frac{2^2 \cdot 3^2}{3^3} = \frac{4}{3}$$

$\boxed{\text{see margin}}$

> Theorem:
>
> $$(ab)^n = a^n b^n$$
>
> For example:
>
> $(2\cdot 3)^2 = 2^2 \cdot 3^2$
>
> Please note, however, that $(a+b)^n$ is generally **not equal** to $a^n + b^n$. In particular $(2+3)^2$ is **not equal** to $2^2 + 3^2$.

Can you now manage Question 2.3:

Simplify: $\dfrac{(3+2)^2}{(2\cdot 5)^3}$ Answer: $\dfrac{1}{40}$

If so, go to Question 2.4. If not:

2.3 Simplify   *Click-Video*

(a) $\dfrac{(2+5)^2}{(2\cdot 3)^3}$ (b) $\dfrac{(2\cdot 3\cdot 4)^2}{(2+3+4)^2}$

If you still can't solve Question 2.3: **Go to the tutoring center.**

Question 2.4

Simplify:

$$\left(\frac{2}{3}\right)^2 + \left(\frac{1}{2}\right)^2$$

The correct answer is $\dfrac{25}{36}$. If you got it, move on to Question 2.5. If not, consider the following example:

EXAMPLE 2.4 Simplify:

$$\left(\frac{1}{2}\right)^3 + \left(\frac{3}{4}\right)^2$$

SOLUTION: One approach:

$$\left(\frac{1}{2}\right)^3 + \left(\frac{3}{4}\right)^2 = \left(\frac{1}{2}\right)\left(\frac{1}{2}\right)\left(\frac{1}{2}\right) + \left(\frac{3}{4}\right)\left(\frac{3}{4}\right)$$

$$= \frac{1}{8} + \frac{9}{16} = \frac{2}{16} + \frac{9}{16} = \frac{11}{16}$$

Another approach:

Theorem:

$$\left(\frac{a}{b}\right)^n = \frac{a^n}{b^n}$$

For example:

$$\left(\frac{1}{2}\right)^3 = \frac{1^3}{2^3} \text{ and } \left(\frac{3}{4}\right)^2 = \frac{3^2}{4^2}$$

$$\left(\frac{1}{2}\right)^3 + \left(\frac{3}{4}\right)^2 \underset{\uparrow}{=} \frac{1^3}{2^3} + \frac{3^2}{4^2} = \frac{1}{8} + \frac{9}{16} = \frac{2}{16} + \frac{9}{16} = \frac{11}{16}$$

$$\boxed{\text{see margin}}$$

Can you now manage Question 2.4:

Simplify: $\left(\frac{2}{3}\right)^2 + \left(\frac{1}{2}\right)^2$

Answer: $\frac{25}{36}$

If so, go to Question 2.5. If not:

2.4 Simplify

Click-Video

(a) $\left(\frac{1}{2}\right)^3 + \left(\frac{2}{3}\right)^2$

(b) $\left(\frac{1}{4}\right)^2 - \left(\frac{3}{2}\right)^3$

If you still can't solve Question 2.4: **Go to the tutoring center.**

Question 2.5

Simplify:

$$\left[\left(1 - \frac{1}{2}\right)^2\right]^3$$

The correct answer is $\frac{1}{64}$. If you got it, move on to Question 2.6. If not, consider the following example:

EXAMPLE 2.5 Simplify:

$$\left[\left(2 - \frac{3}{2}\right)^3\right]^2$$

SOLUTION: One approach:

$$\left[\left(2 - \frac{3}{2}\right)^3\right]^2 = \left[\left(\frac{4}{2} - \frac{3}{2}\right)^3\right]^2 = \left[\left(\frac{4-3}{2}\right)^3\right]^2 = \left[\left(\frac{1}{2}\right)^3\right]^2 = \left[\frac{1}{8}\right]^2 = \frac{1}{64}$$

Theorem:

$$(a^m)^n = a^{m \cdot n}$$

For example:

$$\left[\left(2 - \frac{3}{2}\right)^3\right]^2 = \left(2 - \frac{3}{2}\right)^{3 \cdot 2}$$

Another approach:

$$\left[\left(2 - \frac{3}{2}\right)^3\right]^2 \underset{\uparrow}{=} \left(2 - \frac{3}{2}\right)^6 = \left(\frac{4}{2} - \frac{3}{2}\right)^6 = \left(\frac{1}{2}\right)^6 = \frac{1^6}{2^6} = \frac{1}{64}$$

$$\boxed{\text{see margin}}$$

Can you now manage Question 2.5:

Simplify: $\left[\left(1-\frac{1}{2}\right)^2\right]^3$ Answer: $\frac{1}{64}$

If so, go to Question 2.6. If not:

2.5 Simplify   ***Click-Video***

(a) $\left[\left(\frac{1}{2}-1\right)^2\right]^2$ (b) $\left[\left(\frac{2-\frac{1}{2}}{2}\right)^3\right]^2$

If you still can't solve Question 2.5: **Go to the tutoring center.**

| Question 2.6 |

Simplify:

$$2^2 \cdot 2^3 + \frac{3^5}{3^2}$$

The correct answer is 59. If you got it, move on to Question 2.7. If not, consider the following example:

EXAMPLE 2.6   Simplify:

$$\frac{2^2 \cdot 2^4}{\frac{3^6}{3^4}}$$

SOLUTION: One approach:

$$\frac{2^2 \cdot 2^4}{\frac{3^6}{3^4}} = \frac{4 \cdot 16}{\frac{3^4 \cdot 3^2}{3^4}} = \frac{64}{3^2} = \frac{64}{9}$$

Another approach:

$$\frac{2^2 \cdot 2^4}{\frac{3^6}{3^4}} = \frac{2^{2+4}}{3^{6-4}} = \frac{2^6}{3^2} = \frac{64}{9}$$

see margin

Theorem:

$$a^m \cdot a^n = a^{m+n}$$

For example:

$$2^2 \cdot 2^4 = 2^{2+4} = 2^6$$

Theorem: For $m > n$

$$\frac{a^m}{a^n} = a^{m-n}$$

For example:

$$\frac{3^6}{3^4} = 3^{6-4} = 3^2$$

Can you now manage Question 2.6:

Simplify: $2^2 \cdot 2^3 + \frac{3^5}{3^2}$ Answer: 59

If so, go to Question 2.7. If not

2.6 Simplify   ***Click-Video***

(a) $\frac{-2^2 \cdot 2^4}{\frac{4^3}{4}}$ (b) $\left(\frac{1}{2}\right)^3\left(\frac{1}{2}\right)^2 - \frac{\left(\frac{1}{2}\right)^6}{\left(\frac{1}{2}\right)^3}$

If you still can't solve Question 2.6: **Go to the tutoring center.**

| Question 2.7 | # Simplify: |

$$3^{-2} - 2^3$$

The correct answer is $-\dfrac{71}{9}$. If you got it, move on to Question 2.8. If not, consider the following example:

EXAMPLE 2.7   Simplify:

$$2^{-3} + 2^2 - \frac{1}{4}$$

Definition. For any $a \neq 0$ and any positive integer n:

$$a^{-n} = \frac{1}{a^n}$$

With the above definition in place, we note that the equation

$$\frac{a^m}{a^n} = a^{m-n}$$

holds even if $m < n$. For example:

$$\frac{2^2}{2^5} = 2^{2-5} = 2^{-3}$$

$$\longrightarrow \frac{1}{2^3} \longleftarrow$$

SOLUTION:

$$2^{-3} + 2^2 - \frac{1}{4} \underset{\underset{\boxed{\text{see margin}}}{\uparrow \; 2^3}}{=} \frac{1}{2^3} + 4 - \frac{1}{4} = \frac{1}{8} + 4 - \frac{1}{4} = \frac{1}{8} + \frac{32}{8} - \frac{2}{8}$$

$$= \frac{1+32-2}{8} = \frac{31}{8}$$

Can you now manage Question 2.7:

Simplify: $3^{-2} - 2^3$ Answer: $-\dfrac{71}{9}$

If so, go to Question 2.8. If not:

2.7 Simplify   ***Click-Video***

(a) $3^{-2} + \dfrac{2^2}{9}$ (b) $-2^{-2} + (-2)^2$

If you still can't solve Question 2.7: **Go to the tutoring center.**

| Question 2.8 | # Simplify: |

$$5 \cdot 3^{-2} + \frac{1}{2^{-1}}$$

The correct answer is $\dfrac{23}{9}$. If you got it, move on to Question 2.9. If not, consider the following example:

EXAMPLE 2.8   Simplify:

$$\frac{-3^2}{2^{-3}} - 2^{-1}$$

SOLUTION:

$$\frac{-3^2}{2^{-3}} - 2^{-1} = \frac{-9}{\frac{1}{2^3}} - \frac{1}{2} = \frac{-\frac{9}{1}}{\frac{1}{8}} - \frac{1}{2} = -\frac{9}{1} \cdot \frac{8}{1} - \frac{1}{2} = -72 - \frac{1}{2}$$

$$\boxed{\text{invert and multiply}}$$

$$= -\frac{144}{2} - \frac{1}{2}$$

$$= -\frac{145}{2}$$

Can you now manage Question 2.8:

$$\text{Simplify: } 5 \cdot 3^{-2} + \frac{1}{2^{-1}} \qquad\qquad \text{Answer: } \frac{23}{9}$$

If so, go to Question 2.9. If not:

2.8 Simplify   *Click-Video*

$$\text{(a) } -2 \cdot 3^{-1} + \left(\frac{1}{3}\right)^2 \qquad \text{(b) } \left(\frac{1}{2}\right)^{-2} + 3 \cdot 2^{-2}$$

If you still can't solve Question 2.8: **Go to the tutoring center.**

Question 2.9

Definition. For any $a \neq 0$:

$$a^0 = 1$$

With the above definition in place, we note that the equation

$$\frac{a^m}{a^n} = a^{m-n}$$

holds even if $m = n$. For example:

$$\frac{15^2}{15^2} = 15^{2-2} = 15^0$$

$$\longrightarrow 1 \longleftarrow$$

Simplify:

$$\left(1 - \frac{2}{2^2}\right)^{-2} + 2^0$$

The correct answer is 5. If you did not get it, consider the following example:

EXAMPLE 2.9   Simplify:

$$(2 - 2^{-1})^2 - (15)^0$$

SOLUTION:

$$(2 - 2^{-1})^2 - (15)^0 = \left(2 - \frac{1}{2}\right)^2 - 1 = \left(\frac{4}{2} - \frac{1}{2}\right)^2 - 1$$

$$\boxed{\text{see margin}}$$

$$= \left(\frac{3}{2}\right)^2 - 1 = \frac{9}{4} - 1 = \frac{9}{4} - \frac{4}{4} = \frac{5}{4}$$

Can you now manage Question 2.9:

$$\text{Simplify: } \left(1 - \frac{2}{2^2}\right)^{-2} + 2^0 \qquad\qquad \text{Answer: 5}$$

If not:

2.9 Simplify   *Click-Video*

$$\text{(a) } (2 - 3^{-1})^0 + 3^{-2} \qquad \text{(b) } \left(\frac{2}{1 + 2^{-1}}\right)\left(2 - \frac{3}{2}\right)^0$$

If you still can't solve Question 2.9: **Go to the tutoring center.**

	SUMMARY	

DEFINITIONS	For any number a and any positive integer n:
	$$a^n = \overbrace{a \cdot a \cdots a}^{n\text{-times}}$$
	For any $a \neq 0$: $\quad a^{-n} = \dfrac{1}{a^n} \quad$ and $\quad a^0 = 1$

EXPONENT RULES		
	(i) $\quad a^m a^n = a^{m+n}$	When multiplying add the exponents
	(ii) $\dfrac{a^m}{a^n} = a^{m-n}$	When dividing subtract the exponents
	(iii) $(a^m)^n = (a^n)^m = a^{mn}$	A power of a power: multiply the exponents
	(iv) $(ab)^n = a^n b^n$	A power of a product equals the product of the powers
	(v) $\left(\dfrac{a}{b}\right)^n = \dfrac{a^n}{b^n}$	A power of a quotient equals the quotient of the powers

WARNING	In general, the power of a sum is **NOT** the sum of the powers:
	$(2+3)^2$ is **NOT** equal to $2^2 + 3^2$
	$\quad\quad\quad = 25 \quad\quad\quad\quad\quad\quad\quad = 13$

WARNING	-2^2 is **NOT** equal to 4, it is equal to -4:
	$(-2)^2 = (-2)(-2) = 4 \quad$ while $\quad -2^2 = -(2 \cdot 2) = -4$

ADDITIONAL PROBLEMS

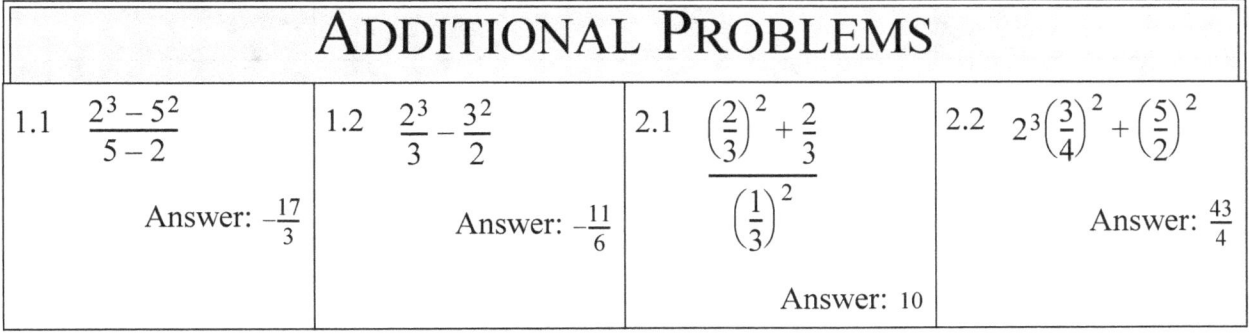

1.1 $\dfrac{2^3 - 5^2}{5 - 2}$	1.2 $\dfrac{2^3}{3} - \dfrac{3^2}{2}$	2.1 $\dfrac{\left(\dfrac{2}{3}\right)^2 + \dfrac{2}{3}}{\left(\dfrac{1}{3}\right)^2}$	2.2 $2^3\left(\dfrac{3}{4}\right)^2 + \left(\dfrac{5}{2}\right)^2$
Answer: $-\dfrac{17}{3}$	Answer: $-\dfrac{11}{6}$	Answer: 10	Answer: $\dfrac{43}{4}$

3.1 $\left(\frac{1}{2}\right)^2 - \left(\frac{1}{2} \cdot \frac{4}{3}\right)^2$ Answer: $-\frac{7}{36}$	3.2 $\dfrac{\left(\frac{2}{3} \cdot \frac{1}{3}\right)^3}{2\left(\frac{2}{3}\right)^4}$ Answer: $\frac{1}{36}$	4.1 $\dfrac{1}{-2^2} + \left(-\frac{1}{2}\right)^2$ Answer: 0	4.2 $\dfrac{1}{-2^3} + \left(-\frac{1}{2}\right)^3$ Answer: $-\frac{1}{4}$
5.1 $-\left(\frac{1}{2} - \frac{1}{3}\right)^2$ Answer: $-\frac{1}{36}$	5.2 $\left(\frac{2}{3} + \frac{1}{2}\right)^2 - \left(\frac{3}{2}\right)^2$ Answer: $-\frac{8}{9}$	5.3 $\left(\frac{2}{3} - \frac{3}{2}\right)^2 + \left(-\frac{3}{2}\right)^2$ Answer: $\frac{53}{18}$	5.4 $\dfrac{\left(\frac{1}{3} - \frac{1}{2}\right)^2}{-\left(\frac{1}{2} \cdot \frac{1}{3}\right)^2}$ Answer: -1
6.1 $\dfrac{\left[\left(1+\frac{1}{2}\right)^3\right]^2}{\left[\left(1+\frac{1}{2}\right)^2\right]^3}$ Answer: 1	6.2 $\left[\left(\frac{1}{2} - 1\right)^2 + \frac{1}{2}\right]^2$ Answer: $\frac{9}{16}$	6.3 $\dfrac{\left[\left(\frac{3}{2}\right)^2 - \frac{3}{2}\right]^2}{1 - \left(\frac{1}{2}\right)^2}$ Answer: $\frac{3}{4}$	6.4 $\left[\dfrac{\left(\frac{1}{3} - \frac{1}{2}\right)^2}{-\left(\frac{1}{2} \cdot \frac{1}{3}\right)^2}\right]^3$ Answer: -1
7.1 $2^{-3} - 2^{-2}$ Answer: $-\frac{1}{8}$	7.2 $\dfrac{\left(1 - \frac{1}{2}\right)^{-2}}{\left(1 - \frac{1}{2}\right)^2}$ Answer: 16	7.3 $\dfrac{3^2 \left(\frac{1}{3}\right)^{-1}}{3^2}$ Answer: 3	7.4 $\left(\dfrac{2^2 + \left(\frac{1}{2}\right)^{-2}}{2}\right)$ Answer: 4
8.1 $\dfrac{3^{-2}}{2^{-3}}$ Answer: $\frac{8}{9}$	8.2 $\left(\frac{1}{2} - 1\right)^{-2} + \frac{1}{2^{-2}}$ Answer: 8	8.3 $\dfrac{\left(\frac{1}{3}\right)^{-1} - \left(\frac{1}{2}\right)^2}{\left(\frac{1}{2}\right)^2}$ Answer: 11	9.1 $\left[\left(1 + \frac{1}{90}\right)^2\right]^0 - 1$ Answer: 0

Sample Test 2
SUPPLEMENT

INTEGER EXPONENTS

It's no big deal to write the expression $3 \cdot 3$, or even the expression $3 \cdot 3 \cdot 3$, but most of us would not want to write out such an expression involving, say fifty 3's. Fortunately, we don't have to:

DEFINITION 2.1   For any positive integer n and any number a:

a RAISED TO THE nth POWER

$$a^n = \underbrace{a \cdot a \cdot \ldots \cdot a}_{n\text{-times}}$$

In the above, a is called the **base**, n is called the **power** or **exponent**, and a^n is called the nth **power of a**, or **a raised to the nth power**.

For example:

$$5^3 = 5 \cdot 5 \cdot 5 = 125 \quad \text{and} \quad 2^5 = 2 \cdot 2 \cdot 2 \cdot 2 \cdot 2 = 32$$

Note that:

$$3^2 \cdot 3^4 = (3 \cdot 3)(3 \cdot 3 \cdot 3 \cdot 3) = 3^6 = 3^{2+4}$$

That:

$$(5^2)^3 = (5^2)(5^2)(5^2) = 5 \cdot 5 \cdot 5 \cdot 5 \cdot 5 \cdot 5 = 5^6 = 5^{2 \cdot 3}$$

And that:

$$(2 \cdot 5)^3 = (2 \cdot 5)(2 \cdot 5)(2 \cdot 5) = (2 \cdot 2 \cdot 2)(5 \cdot 5 \cdot 5) = 2^3 \cdot 5^3$$

Generalizing, we have:

THEOREM 2.1   For any numbers a and b, and any positive integers m and n:

(i) $a^n a^m = a^{n+m}$

(ii) $(a^n)^m = a^{nm}$

(iii) $(ab)^n = a^n b^n$

IN WORDS: (i) When multiplying, add the exponents.

(ii) When taking a power of a power, multiply the exponents.

(iii) The power of a product is the product of the powers.

WARNING: The power of a sum is NOT the sum of the powers.

For example:
$$(3+2)^2 \text{ is NOT equal to } 3^2 + 2^2$$
$$(3+2)^2 = 5^2 = 25 \qquad \textbf{while} \qquad 3^2 + 2^2 = 9 + 4 = 13$$

When it comes to the pecking order of operations, parentheses still rule, but exponents take precedence over multiplication (and division). For example:
$$2^3 \cdot 3 + 4 = 8 \cdot 3 + 4 = 24 + 4 = 28$$
and:
$$[(2+3)^2 + 3(1+4)]^2 = (5^2 + 3 \cdot 5)^2 = (25+15)^2 = 40^2 = 1600$$

EXAMPLE 2.1   (a) Calculate:
$$2^2 + (3+2)^2 + 3 \cdot 2 + 3$$

(b) Simplify:
$$2a^2 + (2a)^2 + a(a+2)$$

SOLUTION: (a) $2^2 + (3+2)^2 + 3 \cdot 2 + 3 = 4 + 5^2 + 6 + 3$
$$= 4 + 25 + 9 = 38$$

The four pieces in $2a^2 + 4a^2 + a^2 + 2a$ are called **terms**. The terms $2a^2, 4a^2$, and a^2, containing the same power of a, are said to be **like terms**, and they can be combined to arrive at the one term $7a^2$. The $7a^2$ and $2a$ are not like terms, and they can not be combined into one term.

(b)
$$2a^2 + (2a)^2 + a(a+2) = 2a^2 + 4a^2 + a^2 + 2a$$
(see margin): $= (2+4+1)a^2 + 2a = 7a^2 + 2a$

CHECK YOUR UNDERSTANDING 2.1

Simplify:

(a) $(1+4)^2 + 3^2 - (2 \cdot 3)^2 + 2^{2+3}$

(b) $\left(\frac{1}{2} - 1\right)^2 + 8 \cdot 2^2 - (2-3)^2 - [-2^2]^3$

(c) $\left(2 + \frac{1}{2}\right)^2 - \left[-1 - \frac{1}{3}\right]^2 + \frac{1}{9}$

(d) $(3a)^2 - 2a^2 + 2^2 a + a(2^2 + a)$

Answers: (a) 30   (b) $\dfrac{381}{4}$ (c) $\dfrac{55}{12}$ (d) $8a^2 + 8a$

Solution: Page 108.

We know that $a^n a^m = a^{n+m}$ (**when you multiply, add exponents**).
To find a similar equation involving $\dfrac{a^n}{a^m}$ we consider the expression $\dfrac{5^7}{5^3}$.
Applying the Cancellation property (page 12) we find that:

$$\frac{5^7}{5^3} = \frac{\cancel{5^3}\,5^4}{\cancel{5^3}} = 5^4 \ (\text{or: } 5^{7-3})$$

Tempting us to say: **when you divide, subtract exponents**.

But if we invoke the above rule to the expression $\dfrac{5^3}{5^7}$ we end up with

a seemingly "meaningless" expression, namely: $\dfrac{5^3}{5^7} = 5^{3-7} = 5^{\overset{?}{-}4}$.

"Meaningless", yes, but not when we breathe meaning into it:

DEFINITION 2.2   For any $a \neq 0$ and any positive integer n:
    NEGATIVE EXPONENT

$$a^{-n} = \frac{1}{a^n}$$

Note how well it works out: $\dfrac{5^3}{5^7} = \dfrac{5^3}{5^3 5^4} = \dfrac{1}{5^4}$ and $5^{-4} = \dfrac{1}{5^4}$.

But there is still a small problem: what are we to do with an expression of the form $\dfrac{5^3}{5^3}$? It is certainly equal to 1, but if we merrily subtract exponents we come up with "5^0" which is another seemingly meaningless expression, until:

DEFINITION 2.3   For any $a \neq 0$
    ZERO EXPONENT $a^0 = 1$

Before moving on to some examples, let's upgrade Theorem 2.1:

THEOREM 2.2   For any numbers a and b, and any integers m and n:

(i) $a^n a^m = a^{n+m}$

(ii) $\dfrac{a^n}{a^m} = a^{n-m} \ (a \neq 0)$

(iii) $(a^n)^m = a^{nm}$

(iv) $(ab)^n = a^n b^n$

(v) $\left(\dfrac{a}{b}\right)^n = \dfrac{a^n}{b^n} \ (b \neq 0)$

EXAMPLE 2.2   (a) Evaluate: $\dfrac{2^{-2}+3}{3-2^{-3}}$

(b) Simplify: $\dfrac{(ab)^2 a^{-3}}{b^{-2} a^5}$, and express your answer without negative exponents.

SOLUTION: There is more than one way to go:

(a) One way: $\dfrac{2^{-2}+3}{3-2^{-3}} = \dfrac{\frac{1}{2^2}+3}{3-\frac{1}{2^3}} = \dfrac{\frac{1}{4}+3}{3-\frac{1}{8}} = \dfrac{\frac{13}{4}}{\frac{23}{8}} = \dfrac{13}{\cancel{4}} \cdot \dfrac{\cancel{8}^{\,2}}{23} = \dfrac{26}{23}$

Another way: $\dfrac{2^{-2}+3}{3-2^{-3}} = \dfrac{\frac{1}{4}+3}{3-\frac{1}{8}} = \dfrac{8\left(\frac{1}{4}+3\right)}{8\left(3-\frac{1}{8}\right)} = \dfrac{2+24}{24-1} = \dfrac{26}{23}$

(b) One way:

when multiplying: add exponents
when dividing: subtract exponents

$\dfrac{(ab)^2 a^{-3}}{b^{-2} a^5} = \dfrac{a^2 b^2 a^{-3}}{b^{-2} a^5} = a^{2-3-5} b^{2-(-2)} = a^{-6} b^4 = \dfrac{b^4}{a^6}$

Another way:

$$\dfrac{(ab)^2 a^{-3}}{b^{-2} a^5} \underset{(*)}{=} \dfrac{(ab)^2 b^2}{a^3 a^5} = \dfrac{a^2 b^2 b^2}{a^8} = \dfrac{b^4}{a^6}$$

Note: A **FACTOR** in the numerator (or denominator) of an expression can be moved to the denominator (or numerator) by changing the sign of its corresponding exponent, as was done in (*) above:

$$\dfrac{(ab)^2 a^{-3}}{b^{-2} a^5} = \dfrac{(ab)^2 b^2}{a^3 a^5}$$

Justification:

$$\dfrac{(ab)^2 a^{-3}}{b^{-2} a^5} = \dfrac{(ab)^2 \left(\frac{1}{a^3}\right)}{\left(\frac{1}{b^2}\right) a^5} = \dfrac{\frac{(ab)^2}{a^3}}{\frac{a^5}{b^2}} = \dfrac{(ab)^2}{a^3} \cdot \dfrac{b^2}{a^5} = \dfrac{(ab)^2 b^2}{a^3 a^5}$$

We again remind you that there is an implicit assumption that the given expression is defined. Here, you are to assume that neither a nor b is zero.

DON'T try this "up-and-down" maneuver with an expression such as $\dfrac{2^{-3} + 3^{-1}}{4^{-1} - 3^{-2}}$. In particular, it is **NOT** equal to $\dfrac{4 - 3^2}{2^3 + 3^1}$. Why not?

Because: $\quad \dfrac{4 - 3^2}{2^3 + 3^1} = \dfrac{4 - 9}{8 + 3} = -\dfrac{5}{11}$,while:

$$\frac{2^{-3} + 3^{-1}}{4^{-1} - 3^{-2}} = \frac{\dfrac{1}{2^3} + \dfrac{1}{3}}{\dfrac{1}{4} - \dfrac{1}{3^2}} = \frac{\dfrac{1}{8} + \dfrac{1}{3}}{\dfrac{1}{4} - \dfrac{1}{9}} = \frac{\dfrac{3 + 8}{24}}{\dfrac{9 - 4}{36}} = \frac{\dfrac{11}{24}}{\dfrac{5}{36}} = \frac{11}{24} \cdot \frac{\overset{3}{\cancel{36}}}{5} = \frac{33}{10}$$

CHECK YOUR UNDERSTANDING 2.2

Simplify:

(a) $\dfrac{2^3 \left(\dfrac{1}{2}\right)^2}{\left(1 + \dfrac{1}{2}\right)^{-1}}$

(b) $\left[2^{-1} + \left(\dfrac{1}{2}\right)^{-2} \right]^{-1}$

(c) $\left[\dfrac{5^{-1} + \left(3 - \dfrac{1}{2}\right)^{-2}}{5\left(\dfrac{1}{3} - \dfrac{5}{9}\right)^{17}} \right]^0 + \dfrac{1}{2}$

(d) $\dfrac{(2a)^{-2}(-b)^2}{a(b)^{-1}}$

Answers: (a) 3   (b) $\dfrac{2}{9}$ (c) $\dfrac{3}{2}$ (d) $\dfrac{b^3}{4a^3}$ (providing $a \neq 0$ and $b \neq 0$)

Solution: Page 108.

Sample Test 3
ALGEBRAIC EXPRESSIONS

Supplement for Sample Test 3 starts on page 43.

Question 3.1

Simplify:

$$\frac{(a^2b)^2}{ab^3}$$

The correct answer is $\dfrac{a^3}{b}$. If you got it, move on to Question 3.2. If not, consider the following example:

EXAMPLE 3.1   Simplify:

$$\frac{4(a^3b)^3}{(2a^2)^3(b^2)^2}$$

SOLUTION: One approach:

$$\frac{4(a^3b)^3}{(2a^2)^3(b^2)^2} \uparrow \frac{4(a^3)^3(b)^3}{2^3a^6b^4} = \frac{4a^9b^3}{2^3a^6b^4} \uparrow \frac{4a^6a^3b^3}{2\cdot4a^6bb^3} = \frac{a^3}{2b}$$

| see margin | | factor and cancel |

Recall that:

$$(ab)^n = a^nb^n$$

and that:

$$(a^n)^m = a^{nm}$$

For example:

$$(2a^2)^3 = 2^3(a^2)^3$$
$$= 2^3a^{2\cdot3}$$
$$= 2^3a^6$$

Another approach, using the properties

$$a^na^m = a^{n+m} \text{ and } \frac{a^n}{a^m} = a^{n-m}:$$

$$\frac{4(a^3b)^3}{(2a^2)^3(b^2)^2} = \frac{2^2a^9b^3}{2^3a^6b^4} = 2^{2-3}a^{9-6}b^{3-4} = 2^{-1}a^3b^{-1} = \frac{a^3}{2b}$$

Can you now manage Question 3.1:

Simplify: $\dfrac{(a^2b)^2}{ab^3}$ Answer: $\dfrac{a^3}{b}$

If so, go to Question 3.2. If not:

3.1 Simplify   ***Click-Video***

(a) $\dfrac{(2a^2b^3c)^2}{a^5bc^2}$ (b) $\left(\dfrac{-x^3yz^2}{xy^{-4}}\right)^3$

If you still can't solve Question 3.9: **Go to the tutoring center.**

Question 3.2	**Simplify:**

$$\frac{x^7 + x^2}{x^2}$$

The correct answer is $x^5 + 1$. If you got it, move on to Question 3.3. If not, consider the following example:

EXAMPLE 3.2   Simplify:

$$\frac{4x^3 - 2x^4}{4x^3}$$

Oh so **WRONG**:

$$\frac{4x^3 - 2x^4}{4x^3}$$ No, No, No!

You can **ONLY** cancel a common **FACTOR**:

$$\frac{a\cancel{c}}{b\cancel{c}} = \frac{a}{b}$$

(providing $c \neq 0$)

SOLUTION: (The answer is **NOT** $1 - 2x^4$ — see margin)

pull out the common factor

$$4x^3 - 2x^4 = 2(2x^3) - x(2x^3) = 2x^3(2-x)$$

$$\frac{4x^3 - 2x^4}{4x^3} \overset{\downarrow}{=} \frac{2x^3(2-x)}{4x^3} = \frac{2x^3(2-x)}{2(2x^3)} = \frac{2-x}{2} \quad \left(\text{or: } 1 - \frac{x}{2}\right)$$

Can you now manage Question 3.2:

Simplify: $\dfrac{x^7 + x^2}{x^2}$ Answer: $x^5 + 1$

If so, go to Question 3.3. If not:

3.2 Simplify   *Click-Video*

(a) $\dfrac{3a^4 + 9a^2}{3a^2}$ (b) $\dfrac{4x^2yz}{2x^2y + 4x^2z}$

If you still can't solve Question 3.2: **Go to the tutoring center.**

Question 3.3	**Simplify:**

$$\left(\frac{\frac{a}{4}}{\frac{a^2}{16}}\right)\left(\frac{a}{16}\right)$$

The correct answer is $\dfrac{1}{4}$. If you got it, move on to Question 3.4. If not, consider the following example:

EXAMPLE 3.3   Simplify:

$$\left(\dfrac{2}{\dfrac{x^3}{4}}\right)\left(\dfrac{\dfrac{x}{2}}{3}\right)$$

SOLUTION:

$$\left(\dfrac{2}{\dfrac{x^3}{4}}\right)\left(\dfrac{\dfrac{x}{2}}{3}\right) = \left(\dfrac{\dfrac{2}{1}}{\dfrac{x^3}{4}}\right)\left(\dfrac{\dfrac{x}{2}}{\dfrac{3}{1}}\right)$$

invert and multiply: $= \left(\dfrac{2}{1}\cdot\dfrac{4}{x^3}\right)\left(\dfrac{x}{2}\cdot\dfrac{1}{3}\right) = \dfrac{\cancel{2}\cdot 4\cdot\cancel{x}}{\cancel{2}\cdot 3\cdot x^2\cdot\cancel{x}} = \dfrac{4}{3x^2}$

Can you now manage Question 3.3:

Simplify: $\left(\dfrac{\dfrac{a}{4}}{\dfrac{a^2}{16}}\right)\left(\dfrac{a}{16}\right)$ Answer: $\dfrac{1}{4}$

If so, go to Question 3.4. If not:

3.3 Simplify   ***Click-Video***

(a) $\dfrac{a^2 b}{2c}\cdot\dfrac{\dfrac{4}{(ab)^2}}{abc}$ (b) $\left(\dfrac{2x^2+6x^3}{\dfrac{2x^2}{3}}\right)\left(\dfrac{1}{2+6x}\right)$

If you still can't solve Question 3.3: **Go to the tutoring center.**

Question 3.4   **Simplify:**

$$\dfrac{2}{a}-\dfrac{1}{ba}+\dfrac{a}{3b}$$

The correct answer is $\dfrac{a^2+6b-3}{3ab}$. If you got it, move on to Question 3.5. If not, consider the following example:

EXAMPLE 3.4   Simplify:

$$\dfrac{a}{2b}+\dfrac{-1}{4ab}-\dfrac{b}{6a}$$

SOLUTION: The first order of business is to multiply the numerator and denominator of each fraction by whatever it takes to express each fraction with a denominator that is the least common denominator: $12ab$:

$$\frac{a}{2b} + \frac{-1}{4ab} - \frac{b}{6a} = \frac{a(6a)}{2b(6a)} + \frac{-1(3)}{4ab(3)} - \frac{b(2b)}{6a(2b)}$$

$$= \frac{6a^2}{12ab} - \frac{3}{12ab} - \frac{2b^2}{12ab} = \frac{6a^2 - 3 - 2b^2}{12ab}$$

Can you now manage Question 3.4:

Simplify: $\dfrac{2}{a} - \dfrac{1}{ab} + \dfrac{a}{3b}$ Answer: $\dfrac{a^2 + 6b - 3}{3ab}$

If so, go to Question 3.5. If not:

3.4 Simplify   *Click-Video*

(a) $\dfrac{ab}{c} + \dfrac{ac}{2b} - \dfrac{bc}{6a}$ (b) $\dfrac{\dfrac{b}{3a} - \dfrac{b}{2a}}{\dfrac{a}{3b} + \dfrac{a}{2b}}$

If you still can't solve Question 3.4: **Go to the tutoring center.**

Question 3.5

Expand:

$$(2x - 3)^2$$

The correct answer is $4x^2 - 12x + 9$. If you got it, move on to Question 3.6. If not, consider the following example:

EXAMPLE 3.5   Expand:

$$(3a + 2b)^2$$

Here is the distributive property:

$a(b + c) = ab + ac$

and here is its pattern:

$\Box(\bigcirc + \triangle) = \Box\bigcirc + \Box\triangle$

For example:

$(3a + 2b)(3a + 2b)$

$= (3a + 2b)3a + (3a + 2b)(2b)$

SOLUTION: One approach:

the distributive property (see margin)

$(3a + 2b)^2 = (3a + 2b)(3a + 2b) = (3a + 2b)3a + (3a + 2b)2b$

$= 9a^2 + 6ba + 6ab + 4b^2$

$= 9a^2 + 12ab + 4b^2$

Another approach using the useful formula:

Using the distributive property:

$(a + b)^2 = (a + b)(a + b)$

$= (a + b)a + (a + b)b$

$= a^2 + ba + ab + b^2$

$= a^2 + 2ab + b^2$

In the same way you can show that:

$(a - b)^2 = a^2 - 2ab + b^2$

square the first term   square the second term

twice the product of the two terms

$$(a + b)^2 = a^2 + 2ab + b^2$$

(see margin)

square the first term   twice the product of the two terms   square the second term

$$(3a + 2b)^2 = (3a)^2 + 2(3a)(2b) + (2b)^2 = 9a^2 + 12ab + 4b^2$$

Can you now manage Question 3.5:

Expand: $(2x-3)^2$ Answer: $4x^2 - 12x + 9$

If so, go to Question 3.6. If not:

3.5 Expand: *Click-Video*

(a) $(2ab + c)^2$ (b) $\left(3x - \dfrac{1}{3}\right)^2$

If you still can't solve Question 3.5: **Go to the tutoring center.**

Question 3.6 | # Multiply:

$$(6a + 5)(2a^2 - 3a + 7)$$

The correct answer is $12a^3 - 8a^2 + 27a + 35$. If you got it, move on to Question 3.7. If not, consider the following example:

EXAMPLE 3.6   Multiply:

$$(3x + 2)(2x^2 - 4x + 3)$$

SOLUTION: One approach (using the distributive property):

$$(3x + 2)(2x^2 - 4x + 3) = (3x + 2)(2x^2) + (3x + 2)(-4x) + (3x + 2)(3)$$

$$= 6x^3 + 4x^2 - 12x^2 - 8x + 9x + 6$$

$$= 6x^3 - 8x^2 + x + 6$$

Another way (the long multiplication method):

$$2x^2 - 4x + 3$$
$$3x + 2$$

multiply $2x^2 - 4x + 3$ by $3x$: $\quad 6x^3 - 12x^2 + 9x$

multiply $2x^2 - 4x + 3$ by 2: $\qquad\quad 4x^2 - 8x + 6$

add: $\quad 6x^3 - 8x^2 + x + 6$

Can you now manage Question 3.6:

Multiply: $(6a + 5)(2a^2 - 3a + 7)$ Answer: $12a^3 - 8a^2 + 27a + 35$

If so, go to Question 3.7. If not:

3.6 Multiply   *Click-Video*
(a) $(3a - 2)(4a^2 + 3a + 1)$ (b) $x^2(x - 1)(2x^2 + x - 5)$

If you still can't solve Question 3.6: **Go to the tutoring center.**

Question 3.7

Factor:

$$25x^2 - 9$$

The correct answer is $(5x + 3)(5x - 3)$. If you got it, move on to Question 3.8. If not, consider the following example:

EXAMPLE 3.7   Factor:

$$9x^2 - 4$$

To establish the difference of two squares formula, just perform the product on the right side of the equation and observe that it does indeed equal $a^2 - b^2$:

$(a + b)(a - b)$
$\quad = (a + b)a + (a + b)(-b)$
$\quad = a^2 + ba - ab - b^2$
$\quad = a^2 - b^2$

SOLUTION: Taking advantage of the difference of two squares formula:

$$(a^2 - b^2) = (a + b)(a - b) \qquad \text{(see margin)}$$

we have:

$$9x^2 - 4 = (3x)^2 - 2^2 = (3x + 2)(3x - 2)$$

Can you now manage Question 3.7:

Factor: $25x^2 - 9$ Answer: $(5x + 3)(5x - 3)$

If so, go to Question 3.8. If not:

3.7 Factor   ***Click-Video***

    (a) $49x^2 - 1$ (b) $x^4 - 16$

If you still can't solve Question 3.7: **Go to the tutoring center.**

Question 3.8

Factor:

$$6x^2 - 11x - 10$$

The correct answer is $(3x + 2)(2x - 5)$. If you got it, move on to Question 3.9. If not, consider the following example:

EXAMPLE 3.8   Factor:

$$5x^2 + 27x - 18$$

SOLUTION: Some trinomials (three terms), such as:

$$5x^2 + 27x - 18$$

can be factored by trial and error. First, consider the "template":

$$(5x \qquad)(x \qquad)$$

which gives you the term $5x^2$. Next, envision pairs of integers in the template whose product is -18, in the hope of finding one for which the middle term, which is the sum of the product of two "outer terms" with the product of the two "inner terms," turns out to be $27x$. In particular, one quickly sees that the first combination below is wrong, while the second is correct:

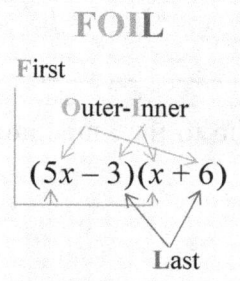

FOIL

First

Outer-Inner

$(5x-3)(x+6)$

Last

$(5x-6)(x+3)$ **NO**

$15x - 6x = 9x$

$(5x-3)(x+6)$ YES

$30x - 3x = 27x$

Thus: $5x^2 + 27x - 18 = (5x-3)(x+6)$

Can you now manage Question 3.8:

Factor: $6x^2 - 11x - 10$ Answer: $(3x+2)(2x-5)$

If so, go to Question 3.9. If not:

3.8 Factor   *Click-Video*

(a) $2x^2 - 7x - 15$ (b) $2x^3 + 6x^2 - 8x$

If you still can't solve Question 3.8: **Go to the tutoring center.**

Question 3.9   Simplify:

$$\frac{3x^2 + 7x + 2}{2x^2 + 5x + 2}$$

The correct answer is $\dfrac{3x+1}{2x+1}$. If you did not get it, consider the following example:

EXAMPLE 3.9   Simplify:

$$\frac{2x^2 + x - 3}{4x^2 + 8x + 3}$$

SOLUTION: Factor and cancel:

$$\frac{2x^2 + x - 3}{4x^2 + 8x + 3} = \frac{(2x+3)(x-1)}{(2x+3)(2x+1)} = \frac{x-1}{2x+1}$$

under the assumption that $2x + 3 \neq 0$

Can you now manage Question 3.9:

Simplify: $\dfrac{3x^2 + 7x + 2}{2x^2 + 5x + 2}$ Answer: $\dfrac{3x+1}{2x+1}$

If not:

3.9 Simplify   *Click-Video*

(a) $\dfrac{x^2 - 4}{3x^2 + x - 10}$ (b) $\dfrac{4x^2 - 14x - 30}{4x^2 + 4x - 3}$

	SUMMARY

| WARNING | You can only cancel a factor that is common to the numerator and the denominator, as with: $$\frac{\cancel{x^2}(3x+5)}{\cancel{x^2}(3x-2)} = \frac{3x+5}{3x-2}$$ You can **NOT** cancel further! Yes, there is a $3x$ in the numerator and denominator of $\frac{3x+5}{3x-2}$, but it is **NOT** a factor — it is **NOT** $3x$ **TIMES** something over $3x$ **TIMES** something. |
| **DIFFERENCE OF TWO SQUARES FORMULA:** | $$(a^2 - b^2) = (a+b)(a-b)$$ |

ADDITIONAL PROBLEMS

1.1 Simplify: $$\frac{b^2c^3}{(2bc)^2}$$ Answer: $\frac{c}{4}$	1.2 Simplify: $$\frac{4x^2y^3}{(2yx^2)^2}$$ Answer: $\frac{y}{x^2}$	1.3 Simplify: $$\frac{(2a^{-2}c)^2}{ac^{-1}}$$ Answer: $\frac{4c^3}{a^5}$	1.4 Simplify: $$\frac{2^{-2}xy^2}{2x^2y^{-1}}$$ Answer: $\frac{y^3}{8x}$
2.1 Simplify: $$\frac{5x^2 - 10x^4}{5x^3}$$ Answer: $\frac{1-2x^2}{x}$	2.2 Simplify: $$\frac{a^2x^2 + a^3x^4}{a^3x^4}$$ Answer: $\frac{1+ax^2}{ax^2}$	2.3 Simplify: $$\frac{a^2b^3}{ab^2 - (ab)^3}$$ Answer: $\frac{ab}{1-a^2b}$	2.4 Simplify: $$\frac{x^2y^3}{(xy)^2 + 2xy}$$ Answer: $\frac{xy^2}{xy+2}$
3.1 Simplify: $$\frac{\frac{ax}{y^2}}{\frac{x^2}{ay}}$$ Answer: $\frac{a^2}{xy}$	3.2 Simplify: $$\frac{\frac{2(-b)^2}{(ab)^3}}{\frac{ab}{b^2}}$$ Answer: $\frac{2}{a^4}$	3.3 Simplify: $$\frac{\frac{yz}{ab^2}}{\frac{yz^2}{a^2b}}$$ Answer: $\frac{a}{zb}$	3.4 Simplify: $$\frac{\frac{a^2b}{(2a^2b)^2}}{a^2}$$ Answer: $\frac{1}{4b}$

3.5 Simplify: $\dfrac{(x^2y)^{-2}}{\dfrac{x^2}{2xy}}$ Answer: $\dfrac{2}{yx^5}$	4.1 Simplify: $\dfrac{2y}{3x}+\dfrac{3x}{2y}$ Answer: $\dfrac{4y^2+9x^2}{6xy}$	4.2 Simplify: $\dfrac{2b^2}{3a}-\dfrac{1}{ab}+\dfrac{2a^2}{3b}$ Answer: $\dfrac{2b^3-3+2a^3}{3ab}$	4.3 Simplify: $\dfrac{\dfrac{3}{ab}-\dfrac{2}{ac}}{\dfrac{2}{abc}}$ Answer: $\dfrac{3c-2b}{2}$
4.4 Simplify: $\dfrac{\dfrac{3}{2xy}+\dfrac{1}{x^2y}}{3x+2}$ Answer: $\dfrac{1}{2x^2y}$	5.1 Multiply: $(2a-b)^2$ Answer: $4a^2-4ab+b^2$	5.2 Multiply: $(x+2)(3x-7)$ Answer: $3x^2-x-14$	5.3 Multiply: $(2x+4y)(-3x+y)$ Answer: $-6x^2-10xy+4y^2$

6.1 Multiply: $(2x+3)(x^2-3x+4)$ Answer: $2x^3-3x^2-x+12$	6.2 Multiply: $(x^2+x-1)(2x-5)$ Answer: $2x^3-3x^2-7x+5$	6.3 Multiply: $(x^2+x-1)(x^2-3x+4)$ Answer: x^4-2x^3+7x-4
7.1 Factor: $(16x^2-25)$ Answer: $(4x+5)(4x-5)$	7.2 Factor: (x^3-49x) Answer: $x(x+7)(x-7)$	7.3 Factor: (x^4-1) Answer: $(x^2+1)(x+1)(x-1)$
8.1 Factor: $(8x^2-10x-3)$ Answer: $(2x-3)(4x+1)$	8.2 Factor: $8x^2+22x+5$ Answer: $(4x+1)(2x+5)$	8.3 Factor: $2x^4+7x^3+3x^2$ Answer: $x^2(2x+1)(x+3)$
9.1 Simplify: $\dfrac{x^2-9}{x^2+x-6}$ Answer: $\dfrac{x-3}{x-2}$	9.2 Simplify: $\dfrac{2x^2+5x+3}{x^2-4x-5}$ Answer: $\dfrac{2x+3}{x-5}$	9.3 Simplify: $\dfrac{2x^3-x^2-15x}{2x^3+9x^2+10x}$ Answer: $\dfrac{x-3}{x+2}$

Sample Test 3
SUPPLEMENT

SIMPLIFYING ALGEBRAIC EXPRESSIONS

An **algebraic expression** is a "meaningful" combination of numbers and variables subjected to operations such as addition, multiplication and powers. You were asked to reduce some algebraic expressions to lowest terms in Supplement 1 [Example 1.1(c), Example 1.2(c), and Example 1.3(d)]. In the next example, we utilize the exponent rules of Supplement 2 (reappearing as Theorem 3.1 below), to simplify other algebraic expressions.

THEOREM 3.1   For any numbers a and b, and any integers m and n:

   (i) $a^n a^m = a^{n+m}$

   (ii) $\dfrac{a^n}{a^m} = a^{n-m} \ (a \neq 0)$

   (iii) $(a^n)^m = a^{nm}$

   (iv) $(ab)^n = a^n b^n$

   (v) $\left(\dfrac{a}{b}\right)^n = \dfrac{a^n}{b^n} \ (b \neq 0)$

Before turning to some specific examples, we recall the cancellation property which already played an important role in our previous development:

THEOREM 3.2   For any number a, and any nonzero numbers b and c:

$$\frac{a\cancel{c}}{b\cancel{c}} = \frac{a}{b}$$

In words: You can **only cancel** a **common nonzero factor** of the numerator and denominator.

For example, while $\dfrac{a^2 b^4}{b^2} = a^2 b^2$ is legitimate (assuming $b \neq 0$), there is absolutely no canceling that can take place in an expression such as $\dfrac{a^2 + b^4}{b^2}$. You can, of course, break the expression into two pieces, and then cancel within one of the pieces:

$$\frac{a^2 + b^4}{b^2} = \frac{a^2}{b^2} + \frac{b^4}{b^2} = \frac{a^2}{b^2} + b^2$$

EXAMPLE 3.1   Simplify (assume no expression is zero):

(a) $\dfrac{a}{3} \cdot \dfrac{15}{-a^2}$ (b) $\dfrac{(2ab)^2 c^3 a^4}{(ab)^{-2}(2c)^3}$

(c) $\dfrac{a(x+y) - a^2(x+y)^2}{a^2(x+y)^5}$ (d) $\dfrac{(1-a)\left(\frac{a}{2}\right)}{\frac{a-1}{2}}$

SOLUTION:

Don't forget that when you see an expression such as $\frac{a}{3} \cdot \frac{15}{-a^2}$, you are to assume that $a \neq 0$.

(a) $\dfrac{a}{3} \cdot \dfrac{15}{-a^2} = \dfrac{\cancel{3} \cdot 5\cancel{a}}{\cancel{3}(-1)\cancel{a} \cdot a} = \dfrac{5}{-a} = -\dfrac{5}{a}$, **OR:** $\dfrac{\cancel{a}}{\cancel{3}} \cdot \dfrac{\cancel{15}^{5}}{-\cancel{a^2}} = \dfrac{5}{-a} = -\dfrac{5}{a}$

(b) $\dfrac{(2ab)^2 c^3 a^4}{(ab)^{-2}(2c)^3} = \dfrac{2^2 a^2 b^2 c^3 a^4}{\dfrac{2^3 c^3}{(ab)^2}} = \dfrac{2^2 a^6 b^2 c^3}{1} \cdot \dfrac{(ab)^2}{2^3 c^3}$

$= \dfrac{2^2 a^6 b^2 \cancel{c^3} a^2 b^2}{\cancel{2^3} \cancel{c^3}} = \dfrac{a^8 b^4}{2}$

canceling common factors

(c) $\dfrac{a(x+y) - a^2(x+y)^2}{a^2(x+y)^5} = \dfrac{\cancel{a}\cancel{(x+y)}[1 - a(x+y)]}{\cancel{a}\cancel{(x+y)}[a(x+y)^4]} = \dfrac{1 - a(x+y)}{a(x+y)^4}$

$2(1-a)\left(\frac{a}{2}\right)$

$= (1-a) \cdot 2 \cdot \frac{a}{2}$

(d) $\dfrac{(1-a)\left(\frac{a}{2}\right)}{\dfrac{a-1}{2}} = \dfrac{2(1-a)\left(\frac{a}{2}\right)}{2 \cdot \dfrac{a-1}{2}} = \dfrac{(1-a)a}{(a-1)}$ *see margin*

$= \dfrac{(1-a)a}{(-1)(1-a)} = \dfrac{a}{-1} = -a$

CHECK YOUR UNDERSTANDING 3.1

Simplify (assume no expression is zero):

(a) $\dfrac{(-2a)^2}{5b} \cdot \dfrac{-b^2}{10} \cdot \dfrac{50}{4b}$ (b) $\dfrac{(-abc)^2}{(ab)^2 + abc}$ (c) $\dfrac{\dfrac{2a+2b}{4}}{\dfrac{a+b}{8}}$

Answers: (a) $-a^2$ (b) $\dfrac{abc^2}{ab+c}$ (c) 4

Answer: Page 109.

EXAMPLE 3.2   Simplify (assume no expression is zero):

(a) $\dfrac{1}{2b} + \dfrac{2}{3a} + \dfrac{-3}{4b} + \dfrac{5}{3}$ (b) $\dfrac{\dfrac{3b}{a+b} - \dfrac{a}{2a+2b}}{\dfrac{4}{a+b}}$

(c) $\dfrac{x(1-z) - x^2(z-1)^2}{x^2 - x^2 z}$

(a) The least common denominator of $\dfrac{1}{2b} + \dfrac{2}{3a} + \dfrac{-3}{4b} + \dfrac{5}{3}$ is easily seen to be $12ab$. So:

$$\dfrac{1}{2b} + \dfrac{2}{3a} + \dfrac{-3}{4b} + \dfrac{5}{3} = \dfrac{1 \cdot 6a}{2b \cdot 6a} + \dfrac{2 \cdot 4b}{3a \cdot 4b} + \dfrac{-3 \cdot 3a}{4b \cdot 3a} + \dfrac{5 \cdot 4ab}{3 \cdot 4ab}$$

$$= \dfrac{6a}{12ab} + \dfrac{8b}{12ab} + \dfrac{-9a}{12ab} + \dfrac{20ab}{12ab}$$

$$= \dfrac{6a + 8b - 9a + 20ab}{12ab} = \dfrac{-3a + 8b + 20ab}{12ab}$$

(b) $\dfrac{\dfrac{3b}{a+b} - \dfrac{a}{2a+2b}}{\dfrac{4}{a+b}} = \dfrac{\dfrac{3b}{a+b} - \dfrac{a}{2(a+b)}}{\dfrac{4}{a+b}} = \dfrac{\dfrac{2(3b)}{2(a+b)} - \dfrac{a}{2(a+b)}}{\dfrac{4}{a+b}}$

$$= \dfrac{\dfrac{6b-a}{2(a+b)}}{\dfrac{4}{a+b}} = \dfrac{6b-a}{2(a+b)} \cdot \dfrac{(a+b)}{4} = \dfrac{6b-a}{8}$$

(c) An important step toward simplifying the expression $\dfrac{x(1-z)-x^2(z-1)^2}{x^2-x^2z}$ is to observe that:

Since $(-a)^n = a^n$ for any **even** integer n, $(z-1)^n = (1-z)^n$ for any **even** integer n.

$$(z-1)^2 = [(-1)(1-z)]^2 = (-1)^2(1-z)^2 = (1-z)^2:$$

$$\frac{x(1-z)-x^2(z-1)^2}{x^2-x^2z} = \frac{x(1-z)-x^2(1-z)^2}{x^2(1-z)}$$

$$= \frac{x(1-z)\cdot 1 - x(1-z)\cdot x(1-z)}{x[x(1-z)]}$$

$$= \frac{[x(1-z)][1-x(1-z)]}{x[x(1-z)]} = \frac{1-x+xz}{x}$$

CHECK YOUR UNDERSTANDING 3.2

Perform the indicated operations and reduce to lowest terms (assume no expression is zero).

(a) $\dfrac{3}{c} + \dfrac{-6}{bc} + \dfrac{a}{2b}$ (b) $\dfrac{3}{2x+6} - \dfrac{x}{x+3} + \dfrac{1}{4}$ (c) $\dfrac{x}{(-x+2)^2} + \dfrac{3}{x-2}$

Answers: (a) $\dfrac{6b-12+ac}{2bc}$ (b) $\dfrac{-3x+9}{4x+12}$ (c) $\dfrac{4x-6}{(x-2)^2}$

Solution: Page 109.

EXAMPLE 3.3   Simplify:

$$\frac{30x(x+2)^3 - 45x(x+2)^2}{10(x+2)^5}$$

SOLUTION: Factor out the largest common factor of the terms in the numerator, $15x(x+2)^2$, and then cancel as much of it as you can with corresponding factors in the denominator:

$$\frac{30x(x+2)^3 - 45x(x+2)^2}{10(x+2)^5} = \frac{15x(x+2)^2[2(x+2)] - 15x(x+2)^2(3)}{10(x+2)^5}$$

$$= \frac{15x(x+2)^2[2(x+2)-3]}{10(x+2)^5}$$

$$= \frac{5(x+2)^2 \cdot 3x[2x+4-3]}{5(x+2)^2 \cdot 2(x+2)^3}$$

$$= \frac{3x(2x+1)}{2(x+2)^3} = \frac{6x^2+3x}{2(x+2)^3}$$

EXAMPLE 3.4 Simplify:

$$\frac{2x^2 - 5x - 3}{x^2 - 9} \cdot \frac{x^2 + 3x}{4x^2 + 4x + 1} - \frac{2x + 1}{x}$$

SOLUTION: The first step is to factor each expression and then cancel wherever possible:

It is important to remember that you can only cancel a non-zero factor that is common to the numerator and the denominator of an expression.

$$\frac{2x^2 - 5x - 3}{x^2 - 9} \cdot \frac{x^2 + 3x}{4x^2 + 4x + 1} - \frac{2x + 1}{x}$$

$$= \frac{(x - 3)(2x + 1)}{(x - 3)(x + 3)} \cdot \frac{x(x + 3)}{(2x + 1)(2x + 1)} - \frac{2x + 1}{x}$$

$$= \frac{x}{2x + 1} - \frac{2x + 1}{x} = \frac{x^2}{x(2x + 1)} - \frac{(2x + 1)^2}{x(2x + 1)}$$

$$= \frac{x^2 - (4x^2 + 4x + 1)}{x(2x + 1)} = \frac{-3x^2 - 4x - 1}{x(2x + 1)} = -\frac{(3x + 1)(x + 1)}{x(2x + 1)}$$

CHECK YOUR UNDERSTANDING 3.3

Simplify:

(a) $\dfrac{6x^2(5x + 2)^4 - 3x^3(5x + 2)^3}{6(-5x - 2)^6}$

(b) $2x(3x - 1)^{-1} + 4x^2(3x - 1)^{-3}$

Answers: (a) $\dfrac{9x^3 + 4x^2}{2(5x + 2)^3}$ (b) $\dfrac{18x^3 - 8x^2 + 2x}{(3x - 1)^3}$

Solutions: Page 109.

Sample Test 4

Supplement for Sample
Test 4 starts on page 59.

EQUATIONS AND INEQUALITIES

Question 4.1

Solve the equation:
$$5x - 3 = 8x + 6$$

The correct answer is $x = -3$. If you got it, move on to Question 4.2. If not, consider the following example:

EXAMPLE 4.1   Solve the equation:
$$4x + 3 = -2x - 9$$

SOLUTION: A TOUCH OF THEORY:

To determine the solution set of $4x + 3 = -2x - 9$ it will first be necessary to rewrite that equation in a more "manageable form." We note that:

Two equations are **equivalent** if they have the same solutions.

It should be rather apparent that a solution of $4x + 3 = -2x - 9$ is also a solution of $4x + 3 + c = -2x - 9 + c$ for any number c, for we have added the same quantity to both sides of the equation. In fact:

Adding (or subtracting) the same number to both sides of an equation results in an equivalent equation.

In particular, the equation:
$$4x + 3 = -2x - 9$$
is equivalent to:
$$4x + 3 - 3 = -2x - 9 - 3$$
$$4x + 0 = -2x - 9 - 3$$
$$4x = -2x - 9 - 3$$

By the same token, the equation:
$$4x + 3 = -2x - 9$$
is equivalent to:
$$4x + 3 + 2x = -2x - 9 + 2x$$
$$4x + 3 + 2x = -9 + 0$$
$$4x + 3 + 2x = -9$$

EFFECT: That 3 which was previously on the left side is now on the right side, **but with its sign changed.**

$$4x + 3 = -2x - 9$$
$$4x = -2x - 9 - 3$$

EFFECT: That $-2x$ which was previously on the right side is now on the left side, **but with its sign changed.**

$$4x + 3 = -2x - 9$$
$$4x + 3 + 2x = -9$$

Upon understanding why it works, you are certainly justified to invoke the following maneuver:

> You may bring over any term from one side of an equation to the other by simply changing its sign.

For example:

$$4x + 3 = -2x - 9$$

bring over and change signs

$$4x + 2x = -3 - 9$$

Then:

$$6x = -12$$

$$\frac{6x}{6} = \frac{-12}{6}$$

$$x = -2$$

Can you now manage Question 4.1:

Solve: $5x - 3 = 8x + 6$ Answer: $x = -3$

If so, go to Question 4.2. If not:

4.1 Solve   *Click-Video*
(a) $-3x - 12 = 5x + 3$ (b) $3(4x + 7) = 2x - 5$

If you still can't solve Question 4.1: **Go to the tutoring center.**

Question 4.2

Solve the equation:

$$\frac{2x}{3} - 4 = 5x + \frac{5}{6}$$

The correct answer is $x = -\dfrac{29}{26}$. If you got it, move on to Question 4.3.

If not, consider the following example:

EXAMPLE 4.2   Solve the equation:

$$\frac{2x}{5} - \frac{1-x}{3} + 1 = -\frac{2x-1}{15}$$

SOLUTION: Our first step is to get rid of all denominators by multiplying through by 15 — the least common denominator of all fractions involved, and then go on from there:

Note: Multiplying (or dividing) both sides of an equation by a non-zero number results in an equivalent equation.

$$\frac{2x}{5} - \frac{1-x}{3} + 1 = -\frac{2x-1}{15}$$

$$15\left(\frac{2x}{5} - \frac{1-x}{3} + 1\right) = 15\left(-\frac{2x-1}{15}\right)$$

Distribute the 15 and then cancel:

$$\overset{3}{\cancel{15}}\left(\frac{2x}{\cancel{5}}\right) - \overset{5}{\cancel{15}}\left(\frac{1-x}{\cancel{3}}\right) + 15 = -\cancel{15}\left(\frac{2x-1}{\cancel{15}}\right)$$

$$3(2x) - 5(1-x) + 15 = -(2x-1)$$

$$6x - 5 + 5x + 15 = -2x + 1$$

Bring the variable terms to the left and the constant terms to the right:

$$6x + 5x + 2x = 1 + 5 - 15$$

$$13x = -9$$

$$x = -\frac{9}{13}$$

Can you now manage Question 4.2:

Solve: $\dfrac{2x}{3} - 4 = 5x + \dfrac{5}{6}$ Answer: $x = -\dfrac{29}{26}$

If so, go to Question 4.3. If not:

4.2 Solve   ***Click-Video***

(a) $2\left(x + \dfrac{1}{3}\right) = \dfrac{x+1}{3}$ (b) $\dfrac{-3x}{5} - \dfrac{x}{2} + 1 = \dfrac{2x+1}{10}$

If you still can't solve Question 4.2: **Go to the tutoring center.**

| Question 4.3 | # Express x in terms of a, given that: |

$$7x - 5a = -10x + 4$$

The correct answer is $x = \dfrac{5a+4}{17}$. If you got it, move on to Question 4.4. If not, consider the following example:

EXAMPLE 4.3   Express x in terms of a, given that:

$$7x - a = 2x + a - 1$$

SOLUTION:

Move all x's to one side and everything else to the other remembering to change the sign of each transposed term:

$$7x - a = 2x + a - 1$$
$$7x - 2x = a - 1 + a$$

Combine like terms: $5x = 2a - 1$

Divide both sides by 5: $x = \dfrac{2a-1}{5}$

Can you now manage Question 4.3:

Express x in terms of a, given that:

$$7x - 5a = -10x + 4$$

$$\text{Answer: } x = \frac{5a + 4}{17}$$

If so, go to Question 4.4. If not:

4.3 Express x in terms of a, given that: ***Click-Video***

(a) $3(2x - a + 1) = x - a - 5$ (b) $\frac{x - 2a}{3} = \frac{x}{6} + a + 1$

If you still can't solve Question 4.3: **Go to the tutoring center.**

| Question 4.4 |

Solve for y in terms of x, if:

$$2x + \frac{y}{3} = -2 + \frac{4}{3}x$$

The correct answer is $y = -2x - 6$. If you got it, move on to Question 4.5. If not, consider the following example:

EXAMPLE 4.4   Solve for y in terms of x, if:

$$2x + \frac{3}{2}y = 4 + \frac{5}{3}x + y$$

SOLUTION: We begin by multiplying both sides of the equation by 6 to clear the denominators, and go on from there:

$$2x + \frac{3}{2}y = 4 + \frac{5}{3}x + y$$

multiply both sides by 6: $12x + 9y = 24 + 10x + 6y$

$$9y - 6y = 24 + 10x - 12x$$

$$3y = -2x + 24$$

$$y = \frac{-2x + 24}{3}$$

Can you now manage Question 4.4:

Solve for y in terms of x, if:

$$2x + \frac{y}{3} = -2 + \frac{4}{3}x$$

$$\text{Answer: } y = -2x - 6$$

If so, go to Question 4.5. If not:

4.4 Solve for y in terms of x, if: ***Click-Video***

(a) $3x + 2(x - y) = y + 4x + 1$ (b) $\frac{x}{3} + \frac{y}{2} = \frac{1}{2}(x + 2y - 4)$

If you still can't solve Question 4.4: **Go to the tutoring center.**

Question 4.5

Solve:

$$5x - 7 < 8x + 3$$

The correct answer is $x > -\dfrac{10}{3}$. If you got it, move on to Question 4.6.

If not, consider the following example:

EXAMPLE 4.5   Solve:

$$3x - 5 < 5x - 7$$

SOLUTION: A TOUCH OF THEORY:

One solves linear inequalities in exactly the same fashion as one solves linear equations, with one notable exception:

WHEN MULTIPLYING OR DIVIDING BOTH SIDES OF AN INEQUALITY BY A **NEGATIVE** QUANTITY, ONE MUST **REVERSE** THE DIRECTION OF THE INEQUALITY SIGN.

If you multiply both sides of the inequality $-2 < 3$ by the positive number 2, then the inequality sign remains as before:

$$-2 < 3$$

multiply by 2: $-4 < 6$

But if you multiply both sides by a negative quantity, then the sense of the inequality is reversed:

$$-2 \le 3$$

multiply by -2: $\not{0}$
$$4 \ge -6$$

To illustrate:

Equation	**Inequality**
$3x - 5 = 5x - 7$	$3x - 5 < 5x - 7$
$3x - 5x = -7 + 5$	$3x - 5x < -7 + 5$
$-2x = -2$	$-2x < -2$
$x = 1$	$\not{0}$ ⟵*reverse* dividing by a negative number
	$x > 1$

Can you now manage Question 4.5:

Solve: $5x - 7 < 8x + 3$ Answer: $x > -\dfrac{10}{3}$

If so, go to Question 4.6. If not:

4.5 Solve: *Click-Video*

(a) $-4x + 3 > 7x + 2$ (b) $2(-5x + 7) \le -12x + 1$

If you still can't solve Question 4.5: **Go to the tutoring center.**

Question 4.6

Solve:

$$\frac{3x}{5} - \frac{2 - x}{3} < \frac{x - 1}{15}$$

The correct answer is $x < \dfrac{9}{13}$. If you got it, move on to Question 4.7. If not, consider the following example:

EXAMPLE 4.6   Solve:

$$\frac{x}{3} - \frac{3x+1}{2} \le \frac{2x-1}{6} + 1$$

SOLUTION: We begin by multiplying both sides of the inequality by 6 to clear the denominators, and then go on from there:

$$\frac{x}{3} - \frac{3x+1}{2} \le \frac{2x-1}{6} + 1$$

$$6\left(\frac{x}{3} - \frac{3x+1}{2}\right) \le 6\left(\frac{2x-1}{6} + 1\right)$$

$$2x - 3(3x+1) \le (2x-1) + 6$$

$$2x - 9x - 3 \le 2x + 5$$

$$2x - 9x - 2x \le 5 + 3$$

$$-9x \le 8$$

reverse dividing by a negative number

$$x \ge -\frac{8}{9}$$

Can you now manage Question 4.6:

Solve: $\dfrac{3x}{5} - \dfrac{2-x}{3} < \dfrac{x-1}{15}$ Answer: $x < \dfrac{9}{13}$

If so, go to Question 4.7. If not:

4.6 Solve: **Click-Video**

(a) $\dfrac{x+1}{3} - \dfrac{2x+1}{6} < \dfrac{x}{2}$ (b) $-\dfrac{1}{3}\left(2x - \dfrac{6x}{5}\right) \ge \dfrac{x+1}{10}$

If you still can't solve Question 4.6: **Go to the tutoring center.**

Question 4.7

Solve:

$$(x-5)(2x+1)(5x-3) = 0$$

The correct answer is $x = 5, x = -\dfrac{1}{2}, x = \dfrac{3}{5}$. If you got it, move on to Question 4.8. If not, consider the following example:

EXAMPLE 4.7   Solve:

$$(x-1)(x+4)(2x+5)(-3x-7) = 0$$

SOLUTION: A TOUCH OF THEORY:
The solution hinges on the following fact:

A PRODUCT IS ZERO IF AND ONLY IF ONE OF THE FACTORS IS ZERO.

In particular, to solve:

$$(x-1)(x+4)(2x+5)(-3x-7) = 0$$

you simply have to determine where each of the four factors is zero:

$x - 1 = 0$	$x + 4 = 0$	$2x + 5 = 0$	$-3x - 7 = 0$
$x = 1$	$x = -4$	$2x = -5$	$-3x = 7$
		$x = -\dfrac{5}{2}$	$x = -\dfrac{7}{3}$

We see that the given equation has exactly four solutions: $1, -4, -\dfrac{5}{2}, -\dfrac{7}{3}$.
Can you now manage Question 4.7:

Solve: $(x-5)(2x+1)(5x-3) = 0$ Answer: $x = 5, x = -\dfrac{1}{2}, x = \dfrac{3}{5}$

If so, go to Question 4.8. If not:

4.7 Solve: *Click-Video*

(a) $-5(x+3)(2x+7)(3x+1) = 0$ (b) $-5x(x+3)(2x+1)^2 = 0$

If you still can't solve Question 4.7: **Go to the tutoring center.**

Question 4.8	**Solve:**

$$4x^2 - 9 = 0$$

The correct answer is $x = -\dfrac{3}{2}, x = \dfrac{3}{2}$. If you got it, move on to Question 4.9. If not, consider the following example:

EXAMPLE 4.8   Solve:

$$9x^3 - x = 0$$

SOLUTION: First factor:

$$9x^3 - x = 0$$
$$x(9x^2 - 1) = 0$$
$$x(3x + 1)(3x - 1) = 0$$

Then use the fact that a product is zero if and only if a factor is zero:

$x = 0$	$3x + 1 = 0$	$3x - 1 = 0$
	$x = -\dfrac{1}{3}$	$x = \dfrac{1}{3}$

We see that the given equation has exactly three solutions: $0, -\dfrac{1}{3}, \dfrac{1}{3}$.

Can you now manage Question 4.8:

Solve: $4x^2 - 9 = 0$ Answer: $x = -\frac{3}{2}, x = \frac{3}{2}$

If so, go to Question 4.9. If not:

4.8 Solve: *Click-Video*

(a) $x^3 - 4x = 0$ (b) $(2x+3)(x^2-25)(x^4+1) = 0$

If you still can't solve Question 4.8: **Go to the tutoring center.**

Question 4.9   Solve:

$$6x^2 - 7x - 3 = 0$$

The correct answer is $x = -\frac{1}{3}, x = \frac{3}{2}$. If you got it, great. If not, consider the following example:

EXAMPLE 4.9   Solve:

$$2x^2 - 5x + 3 = 0$$

SOLUTION:

$$2x^2 - 5x + 3 = 0$$

Factor: $(2x-3)(x-1) = 0$

Set each factor equal to zero: $2x - 3 = 0$ or $x - 1 = 0$

Solve for x: $x = \frac{3}{2}$ or $x = 1$

Can you now manage Question 4.9:

Solve: $6x^2 - 7x - 3 = 0$ Answer: $x = -\frac{1}{3}, x = \frac{3}{2}$

If not:

4.9 Solve: *Click-Video*

(a) $8x^2 - 10x + 3 = 0$ (b) $\frac{x^2}{2} - \frac{x}{2} = 1$

If you still can't solve Question 4.9: **Go to the tutoring center.**

	SUMMARY	
EQUIVALENT EQUATIONS	Two equations are **equivalent** if they have the same solutions. Adding (or subtracting) the same number to both sides of an equation results in an equivalent equation. **Consequently:** You may bring over any term from one side of an equation to the other by simply changing its sign. This will result in an equivalent equation. Multiplying (or dividing) both sides of an equation by a non-zero number results in an equivalent equation.	
LINEAR INEQUALITIES	One solves linear inequalities in exactly the same fashion as one solves linear equations, with one notable exception: WHEN MULTIPLYING OR DIVIDING BOTH SIDES OF AN INEQUALITY BY A **NEGATIVE** QUANTITY, ONE MUST **REVERSE** THE DIRECTION OF THE INEQUALITY SIGN.	
POLYNOMIAL EQUATIONS	If necessary, move all terms to the left side of the equation, so that the right side is zero. If possible factor the polynomial on the left side of the equal sign into a product of linear factors. Utilize: **A PRODUCT IS ZERO IF AND ONLY IF ONE OF THE FACTORS IS ZERO** Then proceed to solve the resulting linear equations stemming from those factors.	

ADDITIONAL PROBLEMS

1.1 Solve: $$3x - 2 = -x - 5$$ Answer: $-\frac{3}{4}$	1.2 Solve: $$-4x - 1 = 5x + 2$$ Answer: $-\frac{1}{3}$	2.1 Solve: $$\frac{x}{5} - \frac{1}{3} = \frac{4}{15}x$$ Answer: -5
2.2 Solve: $$\frac{x}{2} - \frac{2x+1}{3} + 3 = x + \frac{1}{6}$$ Answer: $\frac{15}{7}$	2.3 Solve: $$\frac{1}{3}(2x+1) - 2 = \frac{3x+2}{6}$$ Answer: 12	3.1 Express x in terms of a, if: $$2x - 3a + 1 = x + 3a$$ Answer: $x = 6a - 1$

3.2 Express x in terms of a, if:	4.1 Express y in terms of x, if:	4.2 Express y in terms of x, if:
$$\frac{x-4a}{2}+\frac{1}{4}=x-\frac{a}{4}$$ Answer: $x=\frac{1-7a}{2}$	$$4y-3x+1=2x-4y$$ Answer: $y=\frac{5x-1}{8}$	$$\frac{2x+3y}{5}-x=\frac{y}{10}$$ Answer: $y=\frac{6x}{5}$
5.1 Solve: $$3x+12<7x-14$$ Answer: $x>\frac{13}{2}$	5.2 Solve: $$2+12x\le 7x-2$$ Answer: $x\le -\frac{4}{5}$	6.1 Solve: $$\frac{-3x+1}{4}>\frac{x}{8}+1$$ Answer: $x<-\frac{6}{7}$
6.2 Solve: $$3\left(2x-\frac{1}{5}\right)-x\ge 1$$ Answer: $x\ge \frac{8}{25}$	7.1 Solve: $$(x+1)(2x-3)(-x-5)=0$$ Answer: $-1,\frac{3}{2},-5$	7.2 Solve: $$(-x)\left(\frac{x}{2}-3\right)\left(-\frac{2}{3}x+1\right)=0$$ Answer: $0,6,\frac{3}{2}$
8.1 Solve: $$4x^2-1=0$$ Answer: $-\frac{1}{2},\frac{1}{2}$	8.2 Solve: $$(x^2-9)(4x^2-1)=0$$ Answer: $-3,3,-\frac{1}{2},\frac{1}{2}$	9.1 Solve: $$x^2-3x-10=0$$ Answer: $-2,5$
9.2 Solve: $$9x^2+12x+4=0$$ Answer: $-\frac{2}{3}$	9.2 Solve: $$(x^2-2x-3)(x^2+2x+1)=0$$ Answer: $-1,3$	9.3 Solve: $$x(x^2-49)(2x^2+3x-2)=0$$ Answer: $0,-7,7,-2,\frac{1}{2}$

Sample Test 4
SUPPLEMENT

LINEAR EQUATIONS

The **solution set** of an equation is simply the set of numbers that satisfy the equation (left side of equation equals right side of equation). When solving equations, the following result plays a dominant role:

The fact that you can add (or subtract) the same quantity to both sides of an equation without altering its solution set allows you to move terms from one side of the equation to the other as long as you change the sign of those terms. For example:

$$2x + 3 = 5$$
$$2x + 3 - 3 = 5 - 3$$
$$2x = 5 - 3$$

As you can see that "$+3$" on the left side of the original equation ended up being a "-3" on the right side of the equation.

THEOREM 4.1   Adding (or subtracting) the same quantity to (or from) both sides of an equation, or multiplying (or dividing) both sides of an equation by the same nonzero quantity will not alter the solution set of the equation.

As a consequence of the above Theorem (see margin), we have:

You may bring over any term from one side of an equation to the other by simply changing its sign.

The following equations, in which no variable appears with an exponent greater than 1, are said to be **linear equations** or **first-degree equations:**

$$3x - 4 = -2x + 11$$
$$x + 3y - 2x = 16 - 4y$$
$$2z - 5y - 7x = 3 - x + 8$$

The equation $3x - 4 = -2x + 11$ is a linear equation in the variable x, $x + 3y - 2x = 16 - 4y$ is a linear equation in the variables x and y; and $2z - 5y - 7x = 3 - x + 8$ is a linear equation in the variables x, y, and z.

Several examples on solving linear equations appear in Sample Test 4. We now offer a few more for your consideration.

EXAMPLE 4.1   Solve:
$$3x - x + 2 - 9 = 4x + 6 + x$$

SOLUTION:

$$3x - x + 2 - 9 = 4x + 6 + x$$

Combine like terms on both sides of the equation: $2x - 7 = 5x + 6$

Move all the variable terms to one side of the equation and all the constant terms to the other side, remembering to change signs: $2x - 5x = 6 + 7$

Combine terms once more: $-3x = 13$

Divide both sides by -3: $x = -\dfrac{13}{3}$

Let's show directly that $x = -\frac{13}{3}$ is indeed a solution of the give equation $3x - x + 2 - 9 = 4x + 6 + x$:

$$3\left(-\frac{13}{3}\right) - \left(-\frac{13}{3}\right) + 2 - 9 \stackrel{?}{=} 4\left(-\frac{13}{3}\right) + 6 + \left(-\frac{13}{3}\right)$$

$$-13 + \frac{13}{3} - 7 \stackrel{?}{=} \frac{-52}{3} + 6 - \frac{13}{3}$$

$$-20 + \frac{13}{3} \stackrel{?}{=} 6 - \frac{65}{3}$$

$$-60 + 13 = 18 - 65 \text{ --- yes}$$

It takes longer for us to check our answer than it did for us to solve the equation. Still, make sure you agree with each of our arithmetic steps along the way.

EXAMPLE 4.2   Solve:

$$\frac{2x}{5} - \frac{1-x}{3} + 1 = -\frac{2x-1}{15}$$

SOLUTION: Our first step is to get rid of all denominators by multiplying through by 15 — the least common denominator of all fractions involved, and then go on from there:

$$\frac{2x}{5} - \frac{1-x}{3} + 1 = -\frac{2x-1}{15}$$

$$15\left(\frac{2x}{5} - \frac{1-x}{3} + 1\right) = 15\left(-\frac{2x-1}{15}\right)$$

Distribute the 15 and then cancel:

$$\overset{3}{15}\left(\frac{2x}{5}\right) - \overset{5}{15}\left(\frac{1-x}{3}\right) + 15 = -15\left(\frac{2x-1}{15}\right)$$

$$3(2x) - 5(1-x) + 15 = -(2x-1)$$

$$6x - 5 + 5x + 15 = -2x + 1$$

Bring the variable terms to the left and the constant terms to the right:

$$6x + 5x + 2x = 1 + 5 - 15$$

$$13x = -9$$

$$x = -\frac{9}{13}$$

Hopefully you can follow each step in our solution process. If not, but only after you gave it your **best shot**, you may go to the tutoring center for assistance. But don't ask the tutor to solve the problem for you. Rather, bring your efforts to the tutor and ask: "what am I doing wrong?" You learn little by having someone show you how its done! You learn by trying to do it on your own:

We never understand a thing so well, and make it our own, when we learn it from another as when we have discovered it for ourselves.

Descartes

EXAMPLE 4.3   Solve:

$$\frac{3x-5}{1-\frac{4}{3}} = 2x + \frac{x-1}{2}$$

SOLUTION: We offer one of many procedures that can be used to solve the given equation:

$$\frac{3x - 5}{1 - \frac{4}{3}} = 2x + \frac{x - 1}{2}$$

$$\frac{3x - 5}{-\frac{1}{3}} = \frac{4x}{2} + \frac{x - 1}{2}$$

invert and multiply

$$(3x - 5)(-3) = \frac{4x + x - 1}{2}$$

$$-9x + 15 = \frac{5x - 1}{2}$$

multiply both sides by 2: $-18x + 30 = 5x - 1$

$$-18x - 5x = -1 - 30$$

$$-23x = -31$$

$$x = \frac{31}{23}$$

CHECK YOUR UNDERSTANDING 4.1

Solve:

(a) $3 - 2x + 5 - x = -4x - 2 + 1$ (b) $\frac{-3x}{5} - \frac{x}{2} + 1 = \frac{2x + 1}{10}$

Answers: (a) $x = -9$ (b) $x = \frac{9}{13}$

Solution: Page 110.

CONDITIONAL EQUATION

We point out that a linear equation is said to be **conditional** if it is valid for some value of the variable and not valid for others. The previous equations turned out to be conditional, since each had but one solution.

INCONSISTENT EQUATION

It is possible for a linear equation to have no solution, in which case it is said to be **inconsistent** or a **contradiction**. Such an equation is featured in Example 4.4 below.

IDENTITY

If an equation is valid for all values of the variable, then it is said to be an **identity**. The equation of Example 4.5 below turns out to be an identity.

EXAMPLE 4.4   Solve:

$$3x + 5 - 4x = 5x + 3 - 6x$$

SOLUTION:

Move all the x's to one side and the constants on the other, changing the sign of each transferred term:

$$3x + 5 - 4x = 5x + 3 - 6x$$

$$3x - 4x - 5x + 6x = 3 - 5$$

$$0x = -2$$

Since $0x = 0$ for all x, the equation $0x = -2$ has no solution. It follows that $3x + 5 - 4x = 5x + 3 - 6x$ has no solution.

EXAMPLE 4.5 Solve:

$$4x + 3 - 2x = 5 - x + 3x - 2$$

SOLUTION:

$$4x + 3 - 2x = 5 - x + 3x - 2$$

$$2x + 3 = 2x + 3$$

Clearly, $2x + 3 = 2x + 3$ holds for all values of x, and the given equation is seen to be an identity.

CHECK YOUR UNDERSTANDING 4.2

Determine if the given equation is conditional, or a contradiction, or an identity. In the event that it is conditional, determine its solution and check your answer.

(a) $7x - 5x - 1 = 2x + 3$ (b) $-3x + 7 - x = 2x - 4$

(c) $\dfrac{x-4}{2} = \dfrac{1}{4}(4x - 12) - \dfrac{x}{2} + 1$

Answers: (a) Contradiction (b) Conditional: $x = \dfrac{11}{6}$ (c) Identity

Solution: Page 110.

EXPRESSING ONE VARIABLE IN TERMS OF ANOTHER

Sometimes an equation may contain more than one variable, and you may wish to express one of the variables in terms of the others.

EXAMPLE 4.6 Express x in terms of y, and y in terms of x, given that:

$$\frac{2x+y}{3} = x - \frac{7y}{6} + 1$$

SOLUTION: Let's begin by multiplying both sides by 6:

$$\frac{2x+y}{3} = x - \frac{7y}{6} + 1$$

$$6\left(\frac{2x+y}{3}\right) = 6\left(x - \frac{7y}{6} + 1\right)$$

$$2(2x+y) = 6x - 7y + 6$$

$$4x + 2y = 6x - 7y + 6$$

Then:

Solve for x:

$$4x + 2y = 6x - 7y + 6$$

$$4x - 6x = -7y + 6 - 2y$$

$$-2x = -9y + 6$$

$$x = \frac{-9y+6}{-2} = \frac{9y-6}{2}$$

Solve for y:

$$4x + 2y = 6x - 7y + 6$$

$$2y + 7y = 6x + 6 - 4x$$

$$9y = 2x + 6$$

$$y = \frac{2x+6}{9}$$

CHECK YOUR UNDERSTANDING 4.3

Express y in terms of x, and x in terms of y if:

$$3x + 2(x - y) = y + 4x + 1$$

Answers: $y = \frac{x-1}{3}, x = 3y + 1$

Solution: Page 110.

LINEAR INEQUALITIES

As noted in Sample Test 4, here is the only distinction between solving a linear equation and a linear inequality:

WHEN MULTIPLYING OR DIVIDING BOTH SIDES OF AN INEQUALITY BY A **NEGATIVE** QUANTITY, ONE MUST **REVERSE** THE DIRECTION OF THE INEQUALITY SIGN.

For example, if you multiply both sides of the inequality $\frac{3x}{5} < 4x + 2$ by 5, then you do not reverse the direction of the inequality sign:

$$\frac{3x}{5} < 4x + 2$$

$$5\left(\frac{3x}{5}\right) < 5(4x + 2)$$

$$3x < 20x + 10$$

On the other hand, if you divide both sides of the inequality $-4x < 8$, by -4, then the inequality symbol must be reversed:

$$-4x < 8$$

$$x > \frac{8}{-4} = -2$$

In Example 4.1, we solved the equation:
$3x - x + 2 - 9 = 4x + 6 + x$
The only difference in that solution process and this one, is that when we divide both sides of the inequality by -3, we have to reverse the inequality symbol.

EXAMPLE 4.7 Solve:

$$3x - x + 2 - 9 \geq 4x + 6 + x$$

SOLUTION:

$$3x - x + 2 - 9 \geq 4x + 6 + x$$

Combine like terms on both sides of the equation: $2x - 7 \geq 5x + 6$

Move all the variable terms to one side of the equation and all the constant terms to the other side, remembering to change signs:

$$2x - 5x \geq 6 + 7$$

Combine terms once more: $-3x \geq 13$

Divide both sides by -3: $x \leq -\frac{13}{3}$

CHECK YOUR UNDERSTANDING 4.4

Solve:

$$\frac{2x + 5}{-2} \leq -3x + 1$$

Answers: $x \leq \frac{7}{4}$

Solution: Page 110.

POLYNOMIAL EQUATIONS

A **polynomial of degree n** (in the variable x) is an algebraic expression of the form:

$$a_n x^n + a_{n-1} x^{n-1} + \ldots + a_1 x + a_0$$

where n is a positive integer and a_0 through a_n are numbers, with $a_n \neq 0$.

Every polynomial $p(x) = a_n x^n + a_{n-1} x^{n-1} + \ldots + a_1 x + a_0$ gives rise to a polynomial equation $p(x) = 0$, and the following observation enables you to easily solve such an equation, **providing** you are first able to express $p(x)$ as a product of linear factors.

THEOREM 4.2   A product is zero if and only if one of the factors is zero.

For example, to solve the already nicely factored polynomial equation:

$$x(x-1)(x+4)(2x+5)(-3x-7) = 0$$

you simply have to determine where each of the five factors is zero:

$$
\begin{array}{ccccc}
x = 0 & x - 1 = 0 & x + 4 = 0 & 2x + 5 = 0 & -3x - 7 = 0 \\
 & x = 1 & x = -4 & 2x = -5 & -3x = 7 \\
 & & & x = -\dfrac{5}{2} & x = -\dfrac{7}{3}
\end{array}
$$

We see that the given equation has exactly five solutions: $0, 1, -4, -\dfrac{5}{2}, -\dfrac{7}{3}$.

The above equation was convenient in that it appeared in factored form. This is not the case for the equations in the next example.

EXAMPLE 4.8   Solve the given equation.

(a) $2x^2 - 5x + 3 = 0$

(b) $2x^2 - 20x = x - 2x^2 - 5$

SOLUTION: (a) $\qquad\qquad\qquad 2x^2 - 5x + 3 = 0$

Factor: $\qquad (2x - 3)(x - 1) = 0$

Employ Theorem 4.2: $\quad 2x - 3 = 0 \quad$ or $\quad x - 1 = 0$

Solve for x: $\qquad x = \dfrac{3}{2} \quad$ or $\quad x = 1$

Lets check our answers in the given equation $2x^2 - 5x + 3 = 0$:

$$2\left(\frac{3}{2}\right)^2 - 5\left(\frac{3}{2}\right) + 3 \stackrel{?}{=} 0$$

$$\frac{9}{2} - \frac{15}{2} + 3 \stackrel{?}{=} 0$$

$$-\frac{6}{2} + 3 \stackrel{?}{=} 0 \text{ — yes}$$

$$2(1)^2 - 5(1) + 3 \stackrel{?}{=} 0$$

$$2 - 5 + 3 \stackrel{?}{=} 0 \text{ — yes}$$

(b) To solve the equation $2x^2 - 20x = x - 2x^2 - 5$, begin by bringing all terms to the left, thereby setting it equal to zero:

$$4x^2 - 21x + 5 = 0$$

$$(4x - 1)(x - 5) = 0$$

$$x = \frac{1}{4} \quad \text{or} \quad x = 5$$

Checking in the original equation $2x^2 - 20x = x - 2x^2 - 5$:

$$2\left(\tfrac{1}{4}\right)^2 - 20\left(\tfrac{1}{4}\right) \stackrel{?}{=} \tfrac{1}{4} - 2\left(\tfrac{1}{4}\right)^2 - 5$$

$$\frac{1}{8} - 5 \stackrel{?}{=} \frac{1}{4} - \frac{1}{8} - 5$$

$$1 - 40 \stackrel{?}{=} 2 - 1 - 40 \text{ —yes}$$

$$2(5)^2 - 20(5) \stackrel{?}{=} 5 - 2(5)^2 - 5$$

$$50 - 100 \stackrel{?}{=} 5 - 50 - 5 \text{—yes}$$

CHECK YOUR UNDERSTANDING 4.5

Determine the solution set of the given equation.

(a) $x^2 + x - 6 = 0$ (b) $x^2 - 9 = 0$

(c) $x^2 + 5x - 1 = -x^2 + 4x + 2$

Answers: (a) $x = -3, x = 2$ (b) $x = -3, x = 3$ (c) $x = -\frac{3}{2}, x = 1$

Solution: Page 111.

The "Factoring by grouping" method is used to solve the equation in the following example.

EXAMPLE 4.9 Solve:

$$2x^3 + 3x^2 - 2x - 3 = 0$$

SOLUTION: Staring at the equation $2x^3 + 3x^2 - 2x - 3 = 0$ we observe that if we factor out an x^2 from the first two terms we get $x^2(2x + 3)$, and we can also spot a $(2x + 3)$ factor in the last two terms; leading us to:

$$2x^3 + 3x^2 - 2x - 3 = 0$$
$$x^2(2x + 3) - (2x + 3) = 0$$
$$(2x + 3)(x^2 - 1) = 0$$
$$(2x + 3)(x + 1)(x - 1) = 0$$
$$x = -\frac{3}{2} \quad \text{or} \quad x = -1 \quad \text{or} \quad x = 1$$

CHECK YOUR UNDERSTANDING 4.6

Solve:

$$x^3 + 3x^2 - 4x - 12 = 0$$

Answers: $x = -3, x = -2, x = 2$
Solution: Page 111.

Sample Test 5
LINES AND LINEAR EQUATIONS

Supplement for Sample Test 5 starts on page 83.

Question 5.1

Find the slope of the line passing through the two points $(2, 2)$, $(5, 4)$.

The correct answer is $m = \dfrac{2}{3}$. If you got it, move on to Question 5.2. If not, consider the following example:

> **EXAMPLE 5.1**   Find the slope of the line passing through the two points $(3, -4)$, $(1, 7)$
>
> **SOLUTION:** For any nonvertical line L and any two distinct points (x_1, y_1) and (x_2, y_2) on L, the **slope** of L is that number m given by:
>
> $$m = \frac{y_2 - y_1}{x_2 - x_1} \text{ (see margin)}$$
>
> In this case: $m = \dfrac{7 - (-4)}{1 - 3} = \dfrac{7 + 4}{-2} = -\dfrac{11}{2}$
>
> Or, if you prefer: $m = \dfrac{-4 - 7}{3 - 1} = \dfrac{-11}{2} = -\dfrac{11}{2}$
>
> (You can let $(x_1, y_1) = (3, -4)$, or $(x_1, y_1) = (1, 7)$)

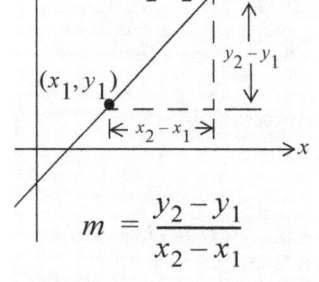

$$m = \frac{y_2 - y_1}{x_2 - x_1}$$

Can you now manage Question 5.1:

> Find the slope of the line passing through $(2, 2)$, $(5, 4)$
>
> Answer: $m = \dfrac{2}{3}$

A line of positive slope is heading up (climbs) — the more positive the slope the steeper the climb.
A line of negative slope is heading down (falls) — the more negative the slope the steeper the fall.

If so, go to Question 5.2. If not:

Click-Video

5.1 Find the slope of the line passing through the two points:

    (a) $(4, -3)$, $(0, -2)$ (b) $(-7, 3)$, $(-2, -5)$

If you still can't solve Question 5.1: **Go to the tutoring center.**

Question 5.2

Find the slope of the line passing through the two points $\left(3, \dfrac{2}{3}\right)$, $\left(\dfrac{5}{2}, -\dfrac{7}{3}\right)$.

The correct answer is $m = 6$. If you got it, move on to Question 5.3. If not, consider the following example:

EXAMPLE 5.2   Find the slope of the line passing through the two points $\left(5, -\frac{7}{3}\right), \left(\frac{1}{2}, 2\right)$

SOLUTION:

$$m = \frac{y_2 - y_1}{x_2 - x_1} = \frac{2 - \left(-\frac{7}{3}\right)}{\frac{1}{2} - 5} = \frac{2 + \frac{7}{3}}{\frac{1}{2} - 5} = \frac{\frac{6}{3} + \frac{7}{3}}{\frac{1}{2} - \frac{10}{2}}$$

$$= \frac{\frac{13}{3}}{-\frac{9}{2}} = -\frac{13}{3} \cdot \frac{2}{9} = -\frac{26}{27}$$

Can you now manage Question 5.2:

Find the slope of the line passing through $\left(3, \frac{2}{3}\right), \left(\frac{5}{2}, -\frac{7}{3}\right)$

Answer: $m = 6$

If so, go to Question 5.3. If not:

Click-Video

5.2 Find the slope of the line passing through the two points:

(a) $\left(\frac{4}{3}, -\frac{1}{2}\right), \left(0, \frac{1}{3}\right)$ (b) $\left(\frac{2}{5}, -1\right), \left(-\frac{1}{5}, -\frac{1}{2}\right)$

If you still can't solve Question 5.2: **Go to the tutoring center.**

| Question 5.3 | **Find the slope-intercept equation of the line of slope $\frac{2}{3}$ and y-intercept 7.** |

The correct answer is $y = \frac{2}{3}x + 7$. If you got it, move on to Question 5.4. If not, consider the following example:

EXAMPLE 5.3   Find the slope-intercept equation of the line of slope 5 and y-intercept $\frac{7}{3}$.

SOLUTION: Some Theory:

Consider the line L of slope m below. Being nonvertical, it must intersect the y-axis at some point $(0, b)$. That number b, where the line intersects the y-axis, is called the **y-intercept** of the line.

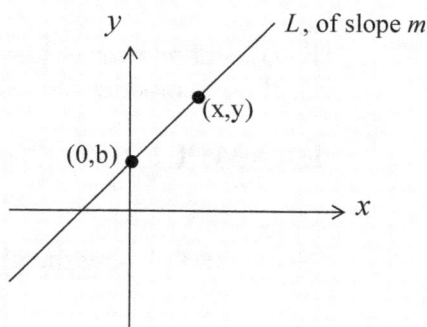

Let (x, y) be any other point on L. From the fact that any two distinct points on a line determine its slope, we have:

$$m = \frac{y - b}{x - 0}$$

multiply both sides of the equation by x: $m = \frac{y - b}{x}$

$\longrightarrow y - b = mx$

add b to both sides of equation: $y = mx + b$

Direct substitution shows that the above equation also holds at the point $(0, b)$.

Bottom line:

The **slope-intercept equation** of the line of slope m and y-intercept b is

$$y = mx + b$$

In particular, here is the slope-intercept equation of the line of slope 5 and y-intercept $\frac{7}{3}$: $y = 5x + \frac{7}{3}$.

Can you now manage Question 5.3:

Find the slope-intercept equation of the line of slope $\frac{2}{3}$ and y-intercept 7.

Answer: $y = \frac{2}{3}x + 7$

If so, go to Question 5.4. If not:

Click-Video

<u>5.3</u> Find the slope-intercept equation of the line:

(a) of slope $-\frac{1}{3}$ and y- intercept $\frac{2}{7}$.   (b) of slope 0 and y- intercept 0.

If you still can't solve Question 5.3: **Go to the tutoring center.**

Question 5.4

Find the slope-intercept equation of the line which contains the points $(3, -4), (5, 6)$.

The correct answer is $y = 5x - 19$. If you got it, move on to Question 5.5. If not, consider the following example:

EXAMPLE 5.4 Find the slope-intercept equation of the line that contains the points $(3, 2)$ and $(-4, 6)$.

SOLUTION: Using the given points, we find that:

$$m = \frac{6-2}{-4-3} = -\frac{4}{7}$$

We now know that the equation is of the form:

$$y = -\frac{4}{7}x + b$$

Since the point $(3, 2)$ is on the line[1], the above equation must hold when 3 is substituted for x and 2 for y, and this enables us to solve for b:

$$2 = -\frac{4}{7}(3) + b$$

$$b = 2 + \frac{12}{7} = \frac{26}{7}$$

Thus:

$$y = -\frac{4}{7}x + \frac{26}{7}$$

To sketch the line in question, simply plot the two given points on the line and then draw the line passing through those points.

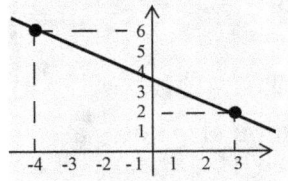

Looking at the above line we see that it has a negative slope, and that its y-intercept is about 4.

Can you now manage Question 5.4:

Find the slope-intercept equation of the line passing through $(3, -4), (5, 6)$

Answer: $y = 5x - 19$

If so, go to Question 5.5 If not:

Click-Video

5.4 Find the slope-intercept equation of the line passing through the two given points:

(a) $(2, -5), (0, 3)$ (b) $\left(\frac{1}{2}, -\frac{1}{3}\right), \left(2, -\frac{1}{2}\right)$

If you still can't solve Question 5.4: **Go to the tutoring center.**

---

1. The other point, $(-4, 6)$ can be used instead of $(3, 2)$, and it would lead to the same result (try it).

| Question 5.5 | **Find the slope and *y*-intercept of the line:** |

$$2x - 3y + 4 = -x + y - 1$$

The correct answer is $m = \dfrac{3}{4}$ and $b = \dfrac{5}{4}$. If you got it, move on to Question 5.6. If not, consider the following example:

EXAMPLE 5.5 Find the slope and *y*-intercept of the line
$$4x + 7y = -y - 3x - 1$$

SOLUTION: Rewrite the equation $4x + 7y = -y - 3x - 1$ in the form $y = mx + b$:

$$4x + 7y = -y - 3x - 1$$
$$7y + y = -4x - 3x - 1$$
$$8y = -7x - 1$$
$$y = \underset{\underset{m}{\uparrow}}{-\frac{7}{8}x} \underset{\underset{b}{\uparrow}}{-\frac{1}{8}}$$

From the above slope-intercept equation, we see that the line has slope $-\dfrac{7}{8}$ and *y*-intercept $-\dfrac{1}{8}$.

Can you now manage Question 5.5:

Find the slope and *y*-intercept of the line $2x - 3y + 4 = -x + y - 1$

Answer: $m = \dfrac{3}{4}$ and $b = \dfrac{5}{4}$

If so, go to Question 5.6. If not:

Click-Video

5.5 Find the slope and *y*-intercept of the line:
(a) $7x = 2y - 4x + 9$ (b) $2(y + 3x) - 1 = 3x - y + 2$

If you still can't solve Question 5.5: **Go to the tutoring center.**

| Question 5.6 | **Find the slope and *y*-intercept of the line:** |

$$\frac{x + 1}{3} = \frac{y}{6} + 1$$

The correct answer is $m = 2$ and $b = -4$. If you got it, move on to Question 5.7. If not, consider the following example:

EXAMPLE 5.6   Find the slope and y-intercept of the line

$$\frac{1}{2}(3x - 4y) + \frac{2x - 3y}{4} = \frac{x}{4} + y + 1$$

SOLUTION: We express the equation in the form $y = mx + b$:

$$\frac{1}{2}(3x - 4y) + \frac{2x - 3y}{4} = \frac{x}{4} + y + 1$$

multiply both sides by 4
to clear denominators:

$$4\left[\frac{1}{2}(3x - 4y) + \frac{2x - 3y}{4}\right] = 4\left[\frac{x}{4} + y + 1\right]$$

$$2(3x - 4y) + (2x - 3y) = x + 4y + 4$$
$$6x - 8y + 2x - 3y = x + 4y + 4$$
$$8x - 11y = x + 4y + 4$$
$$-11y - 4y = -8x + x + 4$$
$$-15y = -7x + 4$$
$$y = \frac{7}{15}x - \frac{4}{15}$$

From the above slope-intercept equation, we see that the line has slope $\frac{7}{15}$ and y-intercept $-\frac{4}{15}$.

Can you now manage Question 5.6:

Find the slope and y-intercept of the line

$$\frac{x + 1}{3} = \frac{y}{6} + 1$$

Answer: $m = 2$ and $b = -4$

If so, go to Question 5.7. If not:

Click-Video

5.6 Find the slope and y-intercept of the line:

(a) $\dfrac{2x - 3y}{5} + \dfrac{1}{3} = \dfrac{1}{3}\left(2x - \dfrac{y}{5}\right) + 1$ (b) $2\left(-x + \dfrac{y}{5}\right) = \dfrac{x + 1}{2 + \dfrac{1}{2}}$

If you still can't solve Question 5.6: **Go to the tutoring center.**

Question 5.7

Find the values of x and y if:
$$3x + 4y = 6 \text{ and } y = 2x + 7$$

The correct answer is $x = -2, y = 3$. If you got it, move on to Question 5.8. If not, consider the following example:

EXAMPLE 5.7   Find the values of x and y if:
$$2x + 3y = 5 \text{ and } y = -x + 1$$

SOLUTION:

Substituting: $y = -x + 1$

in: $2x + 3y = 5$

gives us: $2x + 3(-x + 1) = 5$

$2x - 3x + 3 = 5$ $\rightarrow$ substituting in: $y = -x + 1$

$2x - 3x = 5 - 3$ $\qquad\qquad\qquad y = -(-2) + 1$

$-x = 2$ $\qquad\qquad\qquad\qquad \boxed{y = 3}$

$\boxed{x = -2}$

Let's check our solution in the two given equations. Does the equation $2x + 3y = 5$ hold if $x = -2$ and $y = 3$? Yes:

$$2x + 3y = 2(-2) + 3(3) = -4 + 9 = 5 \text{ Check!}$$

Does the equation $y = -x + 1$ hold if $x = -2$ and $y = 3$? Yes:

$$3 \overset{?}{=} -(-2) + 1 \text{ Yes!}$$

Can you now manage Question 5.7:

Find the values of x and y if $3x + 4y = 6$ and $y = 2x + 7$

Answer: $x = -2, y = 3$

If so, go to Question 5.8. If not:

Click-Video

5.7 Find the values of x and y if:

(a) $2x + 7y = 8$ and $y = x - 4$ (b) $\dfrac{3x + y}{2} = -1$ and $y = \dfrac{x}{3} - \dfrac{4}{3}$

If you still can't solve Question 5.7: **Go to the tutoring center.**

| Question 5.8 | Solve the following system of two equations in two unknowns: |

$$3x + 4y = 6$$
$$x - 2y = -8$$

The correct answer is $x = -2, y = 3$. If you got it, move on to Question 5.6. If not, consider the following example:

To solve a system of two equations in two variables is to determine values of the variables which simultaneously satisfy each equation in the system.

EXAMPLE 5.8   Solve the following system of two equations in two unknowns.

$$(1): \quad -3x + y = 2$$
$$(2): \quad 2x + 2y = 5$$

SOLUTION:

ELIMINATION METHOD: Add (or subtract) a multiple of one equation to a multiple of the other, so as to eliminate one of the variables and arrive at one equation in one unknown:

multiply equation 1 by 2: $2 \times (1): \quad -6x + 2y = 4$

$$(2): \quad 2x + 2y = 5$$

subtract: $-8x \qquad = -1$

$$x = \frac{1}{8}$$

Substituting $x = \frac{1}{8}$ in (1) [we could have chosen (2)], we have:

$$-3\left(\frac{1}{8}\right) + y = 2$$

$$y = 2 + \frac{3}{8} = \frac{19}{8}$$

SUBSTITUTION METHOD:

$$(1): \quad -3x + y = 2$$
$$(2): \quad 2x + 2y = 5$$

Solving for y in (1), we have:

$$y = 3x + 2 \qquad\qquad (*)$$

Substituting this value in (2) yields:

$$2x + 2(3x + 2) = 5$$
$$2x + 6x + 4 = 5$$
$$8x = 1$$

$$x = \frac{1}{8}$$

Returning to (*), we find the corresponding y-value:

$$y = 3 \cdot \frac{1}{8} + 2 = \frac{19}{8}$$

To graph the line $y = 3x + 2$, for example, just plot a couple of its points, say:

$(0, 2)$
↑ ↑
x $y = 3 \cdot 0 + 2 = 2$

and:

$(1, 5)$
↑ ↑
x $y = 3 \cdot 1 + 2 = 5$

and then draw the line passing through those points:

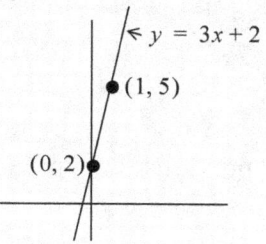

Note: You can take the two given equations and solve each of them for y in terms of x to arrive at the equations of two lines:

(1): $-3x + y = 2 \longrightarrow y = 3x + 2$

(2): $2x + 2y = 5 \longrightarrow y = -x + \frac{5}{2}$

Since every point on the line $y = 3x + 2$ satisfies equation (1), and every point on line $y = -x + \frac{5}{2}$ satisfies equation (2), the point at which the two lines intersect is precisely the solution of the given system of two equations in two unknowns:

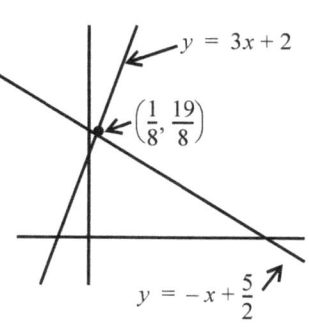

Can you now manage Question 5.8:

Solve: $\left. \begin{array}{r} 3x + 4y = 6 \\ x - 2y = -8 \end{array} \right\}$ Answer: $x = -2, y = 3$

If so, go to Question 5.9. If not:

Click-Video

5.8 Solve:

(a) $\left. \begin{array}{r} 5x + 2y = 3 \\ 2x + y = 1 \end{array} \right\}$ (b) $\left. \begin{array}{r} 3x + 7y = 13 \\ -x + 5y = 3 \end{array} \right\}$

If you still can't solve Question 5.8: **Go to the tutoring center.**

Question 5.9

Solve: $\left. \begin{array}{l} \dfrac{1}{2}(2x + y) = 3x - 1 \\[2mm] \dfrac{x + y}{3} = y - x \end{array} \right\}$

The correct answer is $x = 1, y = 2$. If you did not get it, consider the following example:

EXAMPLE 5.9   Solve:

$$(1): \quad \frac{x+2y}{2} = -\frac{1}{2}$$

$$(2): \quad 3x = \frac{x+y}{4} + 3$$

SOLUTION: The first order of business is to rewrite the two equations in a nicer form:

$$(1): \quad \frac{x+2y}{2} = -\frac{1}{2}$$

multiply both sides by 2: $\quad x + 2y = -1$

$$(2): \quad 3x = \frac{x+y}{4} + 3$$

multiply both sides by 4: $\quad 12x = x + y + 12$

$$12x - x - y = 12$$

$$11x - y = 12$$

At this point, you can use either method of the previous example to solve the system:

$$(3): \quad x + 2y = -1$$
$$(4): \quad 11x - y = 12$$

We elect to go with the elimination method (see margin for the substitution method):

$$(3): \quad x + 2y = -1$$

multiply equation (4) by 2: $\quad 22x - 2y = 24$

add: $\quad 23x \qquad = 23$

$$\boxed{x = 1}$$

substitute 1 for x in equation (3): $\quad 1 + 2y = -1$

$$2y = -2$$

$$\boxed{y = -1}$$

Solve for y in (4):

$$y = 11x - 12$$

Substitute in (3) and solve for x:

$$x + 2(11x - 12) = -1$$
$$x + 22x - 24 = -1$$
$$23x = 23$$
$$\boxed{x = 1}$$

Substitute 1 for x in equation (3) and solve for y:

$$1 + 2y = -1$$
$$2y = -2$$
$$\boxed{y = -1}$$

We check our solution, $x = 1$, $y = -1$, in the original system:

$$(1): \quad \frac{x+2y}{2} = \frac{1+2(-1)}{2} = -\frac{1}{2} \quad \text{Check!}$$

$$(2): 3x \stackrel{?}{=} \frac{x+y}{4} + 3$$

$$3(1) \stackrel{?}{=} \frac{1+(-1)}{4} + 3$$

$$3 \stackrel{?}{=} \frac{0}{4} + 3 \quad \text{Yes!}$$

Can you now manage Question 5.9:

$$\text{Solve:} \quad \left. \begin{array}{l} \dfrac{1}{2}(2x+y) = 3x-1 \\[2mm] \dfrac{x+y}{3} = y-x \end{array} \right\} : \qquad \text{Answer:} \quad x = 1, y = 2$$

If not:

Click-Video

5.9 Solve:

(a) $\left. \begin{array}{l} \dfrac{x}{2}+\dfrac{y}{3} = \dfrac{1}{6} \\[2mm] \dfrac{2x-y}{3} = 1 \end{array} \right\}$

(a) $\left. \begin{array}{l} \dfrac{x+2y}{3} = x-y-\dfrac{4}{3} \\[2mm] -x = \dfrac{3y+1}{2}-\dfrac{5}{2} \end{array} \right\}$

If you still can't solve Question 5.9: **Go to the tutoring center.**

	SUMMARY
SLOPE OF A LINE	For any nonvertical line L and any two distinct points (x_1, y_1) and (x_2, y_2) on L, the **slope** of L is the number m given by: $$m = \frac{y_2 - y_1}{x_2 - x_1} = \frac{\text{change in } y}{\text{change in } x}$$ The more positive the slope, the steeper the climb The the more negative the slope, the steeper the fall
SLOPE-INTERCEPT EQUATION:	A point (x, y) is on the line of slope m and y-intercept b if and only if: $$y = mx + b$$
SOLUTION OF A SYSTEM OF EQUATIONS	To solve a system of equations in several variables is to determine values of the variables which simultaneously satisfy each equation in the system.

ADDITIONAL PROBLEMS

1.1 Find the slope of the line passing through $(1, -2), (7, 7)$.	Answer: $m = \dfrac{3}{2}$
1.2 Find the slope of the line passing through $(-4, -1), (0, -3)$.	Answer: $m = -\dfrac{1}{2}$
2.1 Find the slope of the line passing through $\left(\dfrac{1}{2}, -\dfrac{2}{3}\right), \left(\dfrac{1}{4}, -3\right)$.	Answer: $m = \dfrac{28}{3}$
2.2 Find the slope of the line passing through $\left(2, -\dfrac{1}{3}\right), \left(-\dfrac{1}{3}, \dfrac{1}{2}\right)$.	Answer: $m = -\dfrac{5}{14}$

3.1   Find the slope-intercept equation of the line of slope 7 and y-intercept -3.

Answer: $y = 7x - 3$

3.2   Find the slope-intercept equation of the line of slope 0 and y-intercept 1.   Answer: $y = 1$

4.1   Find the slope-intercept equation of the line that contains the points $(1, 2)$ and $(4, -1)$.

Answer: $y = -x + 3$

4.2   Find the slope-intercept equation of the line that contains the points $(2, 2)$ and $(-2, -1)$.

Answer: $y = \dfrac{3}{4}x + \dfrac{1}{2}$

5.1	Find the slope and y-intercept of the line $4x - 5y = 2y + x - 1$. Answer: $m = \dfrac{3}{7}, b = \dfrac{1}{7}$
5.2	Find the slope and y-intercept of the line $2x - 5 = 3(x + 2y) + 2$. Answer: $m = -\dfrac{1}{6}, b = -\dfrac{7}{6}$
6.1	Find the slope and y-intercept of the line $\dfrac{x-y}{2} = \dfrac{y+x-1}{4}$. Answer: $m = \dfrac{1}{3}, b = \dfrac{1}{3}$
6.2	Find the slope and y-intercept of the line $\dfrac{x-y}{\frac{1}{2}} = \dfrac{y}{2} + x$. Answer: $m = \dfrac{2}{5}, b = 0$
7.1	Find the values of x and y if: $x + 2y = 5$ and $y = 2x$ Answer: $x = 1, y = 2$
7.2	Find the values of x and y if: $3x - 4y = 1$ and $x = y - 1$ Answer: $x = -5, y = -4$
8.1	Solve: $\left.\begin{array}{r} -x + 3y = 2 \\ 4x + 2y = 6 \end{array}\right\}$ Answer: $x = 1, y = 1$
8.2	Solve: $\left.\begin{array}{r} 2x + y = 1 \\ 5x + 2y = 4 \end{array}\right\}$ Answer: $x = 2, y = -3$
9.1	Solve: $\left.\begin{array}{r} \dfrac{x}{2} + \dfrac{y}{3} = \dfrac{5}{6} \\ \dfrac{x+y}{3} = \dfrac{2}{3} \end{array}\right\}$ Answer: $x = 1, y = 1$
9.1	Solve: $\left.\begin{array}{r} \dfrac{x}{2} + y = 3y - \dfrac{1}{2} \\ \dfrac{7}{3}y + \dfrac{x}{2} = -\dfrac{1}{2} \end{array}\right\}$ Answer: $x = -1, y = 0$

Sample Test 5
SUPPLEMENT

THE SLOPE OF A LINE

It is common to associate numerical values with geometrical objects: the area of a rectangle, the circumference of a circle, and so on. The following definition attributes a measure of "steepness" to any non-vertical line in the plane.

DEFINITION 5.1

SLOPE OF A LINE

It can be shown that the slope of a line does not depend on the particular points chosen.

For any nonvertical line L and any two distinct points (x_1, y_1) and (x_2, y_2) on L, we define the **slope** of L to be the number m given by:

$$m = \frac{y_2 - y_1}{x_2 - x_1} \text{ (see Figure 5.1)}$$

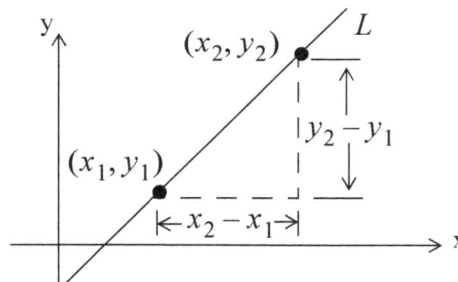

Figure 5.1

The slopes of the lines labeled L_1, L_2, and L_3 in Figure 5.2 are easily determined:

When calculating:
$$m = \frac{y_2 - y_1}{x_2 - x_1}$$
it does not matter which of the two points plays the role of (x_2, y_2).

For example:
$$m_1 = \frac{4 - (-5)}{2 - (-3)} = \frac{9}{5}$$
and:
$$m_1 = \frac{-5 - 4}{-3 - 2} = \frac{9}{5}$$

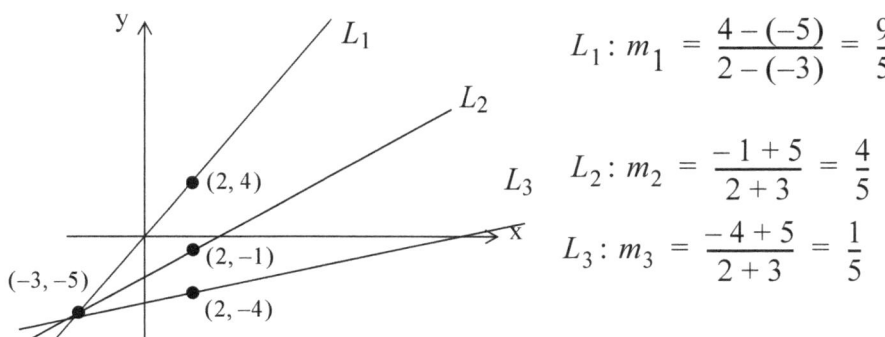

$$L_1 : m_1 = \frac{4 - (-5)}{2 - (-3)} = \frac{9}{5}$$

$$L_2 : m_2 = \frac{-1 + 5}{2 + 3} = \frac{4}{5}$$

$$L_3 : m_3 = \frac{-4 + 5}{2 + 3} = \frac{1}{5}$$

Figure 5.2

The steeper the climb, the more positive the slope.

While lines of positive slope climb as you move to the right, those of negative slope fall:

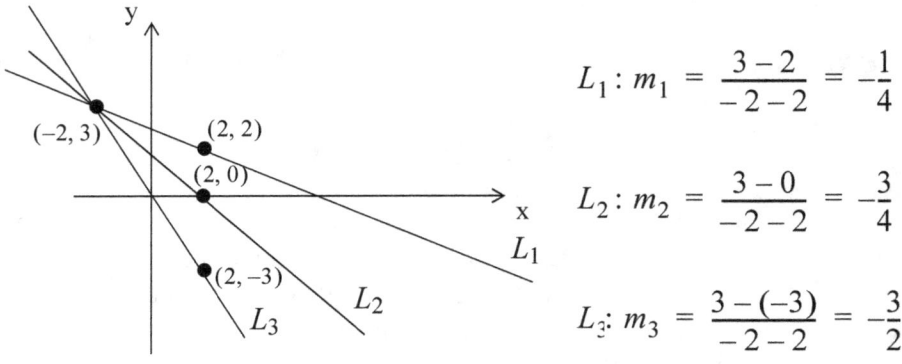

$$L_1 : m_1 \ = \ \frac{3-2}{-2-2} \ = \ -\frac{1}{4}$$

$$L_2 : m_2 \ = \ \frac{3-0}{-2-2} \ = \ -\frac{3}{4}$$

$$L_3 : m_3 \ = \ \frac{3-(-3)}{-2-2} \ = \ -\frac{3}{2}$$

Figure 5.3

The steeper the fall, the more negative the slope

CHECK YOUR UNDERSTANDING 5.1

(a) Determine the slope of the line L that passes through the points $(-1, -6)$ and $(3, 2)$.

(b) If you move 8 units to the right of the point $(-1, -6)$ and then vertically to the line L of part (a), at what point on L do you arrive?

Answers: (a) 2   (b) $(7, 10)$
Solutions: Page 111.

HORIZONTAL AND VERTICAL LINES

If you calculate the slope of any horizontal line you will find that it is zero. Consider, for example, the line L of Figure 5.4(a). Since $(2, 3)$ and $(4, 3)$ are on L:

$$m \ = \ \frac{3-3}{4-2} \ = \ \frac{0}{2} \ = \ 0$$

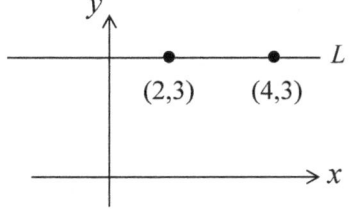

Horizontal lines have slope 0

(a)

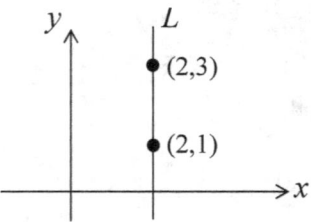

No slope is associated with a vertical line

(b)

Figure 5.4

If you try to calculate the slope of a vertical line you will encounter a problem. Consider, for example, the vertical line L of Figure 5.4(b) which contains the points $(2, 1)$ and $(2, 3)$. An attempt to use these two points (or any other two points on the line) to calculate its "slope" will lead to an undefined expression:

Expressions of the form $\frac{a}{0}$ are undefined (for any a).

$$m = \frac{3 - 1}{2 - 2} = \frac{2}{0} \longleftarrow \text{undefined}$$

SLOPE-INTERCEPT EQUATION OF A LINE

We now develop, for a given non-vertical line L, an equation which is satisfied by every point on L (and only those points). This will afford us an analytical interpretation (the equation) of the line.

Consider the line L of slope m in Figure 5.5. Being nonvertical, it must intersect the y-axis at some point $(0, b)$. That number b, where the line intersects the y-axis, is called the **y-intercept** of the line.

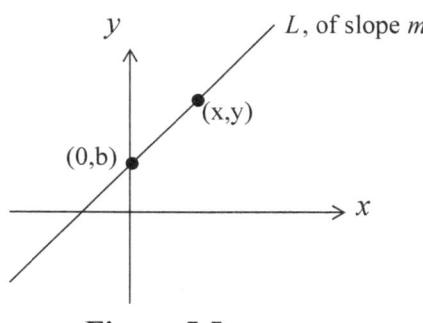

Figure 5.5

Let (x, y) be any other point on L. From the fact that any two distinct points on a line determine its slope, we have:

$$m = \frac{y - b}{x - 0}$$

multiply both sides of the equation by x:

$$m = \frac{y - b}{x}$$

$$y - b = mx$$

add b to both sides of equation: $\quad y = mx + b$

Direct substitution shows that the above equation also holds at the point $(0, b)$; thus:

THEOREM 5.1 A point (x, y) is on the line of slope m and y-intercept b if and only if its coordinates satisfy
Slope-Intercept
Equation of a line the equation $y = mx + b$.

EXAMPLE 5.1 Find the equation of the line that contains the points $(3, 2)$ and $(-4, 6)$.

SOLUTION: Using the given points, we find that:

$$m = \frac{6-2}{-4-3} = -\frac{4}{7}$$

We now know that the equation is of the form:

$$y = -\frac{4}{7}x + b$$

Since the point $(3, 2)$ is on the line, the above equation must hold when 3 is substituted for x and 2 for y, and this enables us to solve for b:

$$2 = -\frac{4}{7}(3) + b$$

The other point, $(-4, 6)$ can be used instead of $(3, 2)$, and it would lead to the same result (try it).

$$b = 2 + \frac{12}{7} = \frac{26}{7}$$

Thus:

$$y = -\frac{4}{7}x + \frac{26}{7}$$

CHECK YOUR UNDERSTANDING 5.2

Find the equation of the line passing through the points $(-3, 5)$ and $(4, -6)$.

Answer: $y = -\frac{11}{7}x + \frac{2}{7}$

Solution: Page 111.

TWO LINEAR EQUATIONS IN TWO UNKNOWNS

Our next concern is with the solution of a system of two linear equations in two unknowns.

$$\begin{matrix} (1): & 2x + 3y = -3 \\ (2): & x + 4y = -9 \end{matrix} \Big\}$$

is an example of such a system, and a solution will consist of a value of x and of y for which **both** of the equations in the system are satisfied. For example, $(x = 3, y = -3)$ is seen to be a solution of the system:

$$2(3) + 3(-3) = -3 \quad \text{Yes}$$
$$3 + 4(-3) = -9 \quad \text{Yes}$$

CHECK YOUR UNDERSTANDING 5.3

Indicate True or False.

(a) $(x = -1, y = 3)$ is a solution of $\left.\begin{array}{r} 2x - 4y = 10 \\ -x + 5y = -8 \end{array}\right\}$

(b) $(x = 3, y = -1)$ is a solution of the system in (a)

(a) False   (b) True

Solution: Page 111.

We shall develop two methods for solving two equations in two unknowns. The first of these methods is called the **substitution method**, and it works as follows:

To determine the values of x and of y for which the two equations

$$(1): \quad 2x + 3y = -3$$
$$(2): \quad x + 4y = -9$$

are simultaneously satisfied, we look to one or the other of the equations in order to express either x in terms of y, or y in terms of x, whichever appears easier. Now, it is quite easy to express x in terms of y from equation (2):

$$(3): \quad x = -9 - 4y$$

Substituting this expression for x in equation (1), we obtain the following linear equation in the *one* unknown, y:

$$2(-9 - 4y) + 3y = -3$$

which is easily solved:

$$-18 - 8y + 3y = -3$$
$$-8y + 3y = -3 + 18$$
$$-5y = 15$$
$$y = -3$$

Next, substitute the value $y = -3$ in equation (3) to obtain

$$x = -9 - 4(-3)$$
$$x = 3$$

At this point, we know that $(x = 3, y = -3)$ is the only possible solution of the system

$$(1): \quad 2x + 3y = -3$$
$$(2): \quad x + 4y = -9$$

Here is another example for your consideration:

EXAMPLE 5.2 Solve:

$$2x - 3y + 1 = 5x - 4y + 3$$
$$x + y = -x - y + 5$$

SOLUTION: Begin by moving and combining all variables on one side of the equation and all constant terms on the other side of each of the two equations, being careful to change the sign of any term that is moved from one side of an equation to the other, to obtain:

$$2x - 3y - 5x + 4y = 3 - 1$$
$$x + y + x + y = 5$$
Or:
$$(1): \quad -3x + y = 2$$
$$(2): \quad 2x + 2y = 5$$

From (1) we have:

$$(3): \quad y = 2 + 3x$$

Substituting in (2):

$$2x + 2(2 + 3x) = 5$$
$$2x + 4 + 6x = 5$$
$$8x = 1$$
$$x = \frac{1}{8}$$

Substituting in (3):

$$y = 2 + 3\left(\frac{1}{8}\right) = \frac{16}{8} + \frac{3}{8} = \frac{19}{8}$$

We leave it for you to verify that $(x = \frac{1}{8}, y = \frac{19}{8})$ satisfies both of the given equations.

CHECK YOUR UNDERSTANDING 5.4

Solve:

(a)
$$2x - 7y = 18$$
$$x + 5y = -8$$

(b)
$$\frac{y + x}{3} + 3y = \frac{1}{6}$$
$$\frac{2x}{5} - \frac{y + 1}{10} = \frac{1}{10}$$

Answer: (a) $(x = 2, y = -2)$ (b) $(x = \frac{1}{2}, y = 0)$

Solution: Page 112.

We now turn to the **elimination method** for solving two equations in two unknowns. Here is how it works:

If a particular x and y is a solution of the system:

$$\left.\begin{array}{ll} (1): & 3x + 7y = -11 \\ (2): & -3x + 2y = -7 \end{array}\right\}$$

then the expression $-3x + 2y$ on the left side of equation (2) equals the number -7; and by adding $-3x + 2y$ to the left side of equation (1), and -7 to the right side of equation (1), we end up with an equation in the one variable, y:

$$\begin{array}{lr} (1): & 3x + 7y = -11 \\ (2): & \underline{-3x + 2y = -7} \\ (1) + (2): & 9y = -18 \\ & y = -2 \end{array}$$

Upon substituting -2 for y in equation (1) [we could have chosen (2)], we obtain:

$$\begin{array}{rcl} 3x + 7(-2) &=& -11 \\ 3x - 14 &=& -11 \\ 3x &=& 3 \\ x &=& 1 \end{array}$$

We have shown that $(x = 1, y = -2)$ is the only possible solution of the given system. Let's check it out:

$$\begin{array}{ccc} 3x + 7y = -11 & x = 1 & 3(1) + 7(-2) = -11 \quad \text{Yes} \\ -3x + 2y = -7 & y = -2 & -3(1) + 2(-2) = -7 \quad \text{Yes} \end{array}$$

EXAMPLE 5.3   Solve:.

$$\left.\begin{array}{ll} (1): & 2x - 3y = 10 \\ (2): & 3x - 11y = 28 \end{array}\right\}$$

SOLUTION: Adding or subtracting equation (2) from equation (1) will not result in the elimination of either of the variables x or y. However, if we multiply *both sides* of equation (1) by 3 and both sides of equation (2) by -2 and add, the x-term drops out:

$$\begin{array}{lr} 3 \times (1): & 6x - 9y = 30 \\ -2 \times (2): & \underline{-6x + 22y = -56} \\ \text{add:} & 13y = -26 \\ & y = -2 \end{array}$$

Substituting in (1) yields:

$$2x - 3(-2) = 10$$
$$2x + 6 = 10$$
$$2x = 4$$
$$x = 2$$

We leave it for you to verify that $(x = 2, y = -2)$ satisfies both of the given equations.

CHECK YOUR UNDERSTANDING 5.5

Solve:

(a) $\left.\begin{array}{l} 4x + 5y = -17 \\ 4x - 2y = -10 \end{array}\right\}$

(b) $\left.\begin{array}{l} 2x + 3y = \dfrac{3}{2} - x - 2y \\ -6x - 9y = -3 \end{array}\right\}$

Answers: (a) $(x = -3, y = -1)$ (b) $(x = \frac{1}{2}, y = 0)$

Solution: Page 112.

Sample Test 6

ODDS AND ENDS

Supplement for Sample Test 6 starts on page 101.

Question 6.1

Determine $f\left(\dfrac{2}{3}\right)$ **if** $f(x) = \dfrac{x^2 - 1}{3x}$.

The correct answer is $-\dfrac{5}{18}$. If you got it, move on to Question 6.2. If not, consider the following example:

EXAMPLE 6.1   Determine $f\left(\dfrac{3}{2}\right)$ if $f(x) = \dfrac{2x^2 + x}{x - 2}$.

SOLUTION: We are asked to evaluate the given function at $x = \dfrac{3}{2}$. To do so, we simply replace x everywhere in $f(x) = \dfrac{2x^2 + x}{x - 2}$ with $\dfrac{3}{2}$:

$$f\left(\frac{3}{2}\right) = \frac{2\left(\frac{3}{2}\right)^2 + \frac{3}{2}}{\frac{3}{2} - 2} = \frac{2\left(\frac{9}{4}\right) + \frac{3}{2}}{\frac{3}{2} - \frac{4}{2}} = \frac{\frac{9}{2} + \frac{3}{2}}{\frac{-1}{2}} = \frac{\frac{12}{2}}{\frac{-1}{2}} = \frac{6}{1} \cdot \frac{2}{-1} = -12$$

Can you now manage Question 6.1:

Determine $f\left(\dfrac{2}{3}\right)$ if $f(x) = \dfrac{x^2 - 1}{3x}$

Answer: $-\dfrac{5}{18}$

If so, go to Question 6.2. If not:

6.1 Determine: *Click-Video*

(a) $f\left(\dfrac{3}{4}\right)$ if $f(x) = 2x^2 - x - 1$ (b) $g\left(-\dfrac{1}{2}\right)$ if $g(x) = \dfrac{-x}{(x + 1)^2}$

If you still can't solve Question 6.1: **Go to the tutoring center.**

Question 6.2

Determine $f(a + 2)$ **if** $f(x) = \dfrac{x^2 + 1}{x + 2}$.

The correct answer is $\dfrac{a^2 + 4a + 5}{a + 4}$. If you got it, move on to Question 6.3. If not, consider the following example:

EXAMPLE 6.2   Determine $f(c-3)$ if $f(x) = \dfrac{x^2-4}{x-5}$.

$$f(x) = \frac{x^2-4}{x-5}$$

$$f\,\Box = \frac{\Box^2 - 4}{\Box - 5}$$

$$f\,\boxed{c-3} = \frac{\boxed{c-3}^2 - 4}{\boxed{c-3} - 5}$$

SOLUTION: It is important for you to know that the variable x is a placeholder; a "box" that can hold any meaningful expression. In particular, to evaluate the function at $c-3$, you put $c-3$ in the box (see margin):

$$\overset{\displaystyle (a-b)^2 = a^2 - 2ab + b^2}{}$$

$$f(c-3) = \frac{(c-3)^2 - 4}{(c-3)-5} = \frac{c^2 - 6c + 9 - 4}{c-3-5} = \frac{c^2 - 6c + 5}{c-8}$$

NOTE: In the expression

$$f(c-3) = \frac{c^2 - 6c + 5}{c-8} \qquad (*)$$

Why can't you have a zero in the denominator? Well $\frac{15}{3} = 5$ since 3 times 5 is 15. Fine, but " $\frac{15}{0}$ " won't do, since no number times 0 is 15.

Neither is the expression " $\frac{0}{0}$ " meaningful, for in this situation all numbers would "work." For example, if you like, you can say that $\frac{0}{0} = 199$ since 0 times 199 is certainly 0.
Bottom line A denominator cannot be zero! (Note, however, that if $a \neq 0$, $\frac{0}{a} = 0$, since a times 0 is 0.

c can be any number **except** 8. Why? Because, if you substitute 8 for c you end up with a **zero in the denominator** of (*), and such an expression is **undefined** (see margin). A zero in the numerator is okay, as long as the denominator is not zero. In particular c can be 1 in (*), since

$$f(1-3) = \frac{1^2 - 6 \cdot 1 + 5}{1-8} = \frac{0}{-7} = 0$$

Can you now manage Question 6.2:

Determine $f(a+2)$ if $f(x) = \dfrac{x^2+1}{x+2}$ Answer: $\dfrac{a^2+4a+5}{a+4}$

If so, go to Question 6.3. If not:

<u>6.2</u> Determine: ***Click-Video***

(a) $f(a+1)$ if $f(x) = \dfrac{x-2}{x^2}$ (b) $g(x+h)$ if $g(x) = x^2 - x + 1$

If you still can't solve Question 6.2: **Go to the tutoring center.**

$\boxed{\text{Question 6.3}}$ # What is 34% of 18?

The correct answer is $\dfrac{153}{25}$ or 6.12 . If you got it, move on to Question 6.4. If not, consider the following example:

EXAMPLE 6.3   What is 22% of 90?

SOLUTION: The first thing you have to know is that for any number a, $a\%$ is defined to be the number $\dfrac{a}{100}$. In particular: $22\% = \dfrac{22}{100}$. You also need to know that, in mathematics, the word "of" is often used to denote multiplication. For example $\dfrac{1}{3}$ of 21 is 7, since $\dfrac{1}{3} \cdot 21 = 7$.

Let's now translate the given question into a mathematical equation, substituting the variable x for the "**What:**"

$$\text{What is } 22\% \text{ of } 90$$

$$x = \frac{22}{100} \cdot 90 = \frac{11}{50} \cdot 90 = \frac{11 \cdot 9}{5} = \frac{99}{5} \text{ or } 19.8$$

$$\text{Answer: } \frac{99}{5} = 19.8 \text{ is } 22\% \text{ of } 90$$

Can you now manage Question 6.3:

What is 34% of 18?

Answer: $\dfrac{153}{25}$ or 6.12

If so, go to Question 6.4. If not:

Click-Video

6.3 (a) What is 120% of 80?   (b) What is 7% of $\dfrac{3}{5}$?

If you still can't solve Question 6.3: **Go to the tutoring center.**

Question 6.4

15 is what percent of 50?

The correct answer is 30%. If you got it, move on to Question 6.5. If not, consider the following example:

EXAMPLE 6.4   27 is what percent of 18?

SOLUTION: It's all in the translation:

$$27 \text{ is } \textbf{what percent } \text{of } 18$$

$$27 = \frac{x}{100} \cdot 18$$

$$2700 = 18x$$

$$x = \frac{2700}{18} = 150 \qquad \text{Answer: 27 is } 150\% \text{ of } 18.$$

Can you now manage Question 6.4?:

15 is what percent of 50?   Answer: 30%

If so, go to Question 6.5. If not:

Click-Video

6.4 (a) 12 is what percent of 4?   (b) 20 is what percent of $\frac{1}{2}$?

If you still can't solve Question 6.4: **Go to the tutoring center.**

Question 6.5

60 is 20 percent of what?

The correct answer is 300. If you got it, move on to Question 6.6. If not, consider the following example:

EXAMPLE 6.5   15 is 75 percent of what?

SOLUTION: 15 is 75% of what

$$15 = \frac{75}{100} \cdot x$$

$$1500 = 75x$$

$$x = \frac{1500}{75} = 20 \qquad \text{Answer: 15 is 75\% of 20.}$$

Can you now manage Question 6.5:

60 is 20% of what?   Answer: 300

If so, go to question 6.6. If not:

Click-Video

6.5 (a) 5 is 20 percent of what?   (b) 120 is 15 percent of what?

If you still can't solve Question 6.5: **Go to the tutoring center.**

Question 6.6

Evaluate:

$$\frac{\sqrt{25} - 9^{\frac{1}{2}}}{2^3}$$

The correct answer is $\frac{1}{4}$. If you got it, move on to Question 6.7. If not, consider the following example:

EXAMPLE 6.6 Evaluate:

$$\frac{-16^{\frac{1}{2}} + \sqrt{49}}{\frac{1}{3}}$$

For $a \geq 0$, the expressions $\sqrt{a}$ or $a^{\frac{1}{2}}$ is called the **principal square root of a,** and is defined to be that **non-negative** number which when squared equals a. For example:

$\sqrt{49} = 7$ since $7^2 = 49$
(and 7 is non-negative)
and

$16^{\frac{1}{2}} = 4$ since $4^2 = 16$
(and 4 is non-negative)

Note: Since the square of any number cannot be negative, expressions such as $\sqrt{-49}$ and $(-16)^{1/2}$ are not defined (in the real number system).

SOLUTION:

$$\frac{-16^{\frac{1}{2}} + \sqrt{49}}{\frac{1}{3}} = \underset{\text{see margin}}{\underset{\uparrow}{\frac{-4 + 7}{\frac{1}{3}}}} = \frac{3}{\frac{1}{3}} = \frac{\frac{3}{1}}{\frac{1}{3}} = \frac{3}{1} \cdot \frac{3}{1} = 9$$

Can you now manage Question 6.6:

Evaluate: $\dfrac{\sqrt{25} - 9^{\frac{1}{2}}}{2^3}$

Answer: $\frac{1}{4}$

If so, go to Question 6.7. If not:

6.6 Evaluate: *Click-Video*

(a) $\dfrac{-\sqrt{9} + \sqrt{36}}{-\sqrt{49}}$ (b) $\dfrac{\left(\frac{16}{25}\right)^{\frac{1}{2}} - 4^{-\frac{1}{2}}}{(16)^{\frac{1}{2}}}$

If you still can't solve Question 6.6: **Go to the tutoring center.**

| Question 6.7 | **Simplify:**

$$\frac{2}{\sqrt{3}} - \sqrt{\frac{2}{5}}$$

(Answer is not to contain any square root in the denominator.)

The correct answer is $\dfrac{10\sqrt{3} - 3\sqrt{10}}{15}$. If you got it, move on to Question 6.8. If not, consider the following example:

EXAMPLE 6.7 Simplify:

$$\frac{5}{\sqrt{2}} + \sqrt{\frac{2}{3}}$$

Recall that for $a \geq 0$, $\sqrt{a}$ is defined to be that number which when squared equals a. In particular $(\sqrt{2})^2 = 2$,

$$\sqrt{\frac{2}{3}} = \left(\frac{2}{3}\right)^{\frac{1}{2}} = \frac{2^{\frac{1}{2}}}{3^{\frac{1}{2}}} = \frac{\sqrt{2}}{\sqrt{3}}$$

| power of a quotient is the quotient of the powers |

and:

$$\sqrt{2} \cdot \sqrt{3} = 2^{\frac{1}{2}} \cdot 3^{\frac{1}{2}}$$

$$= (2 \cdot 3)^{\frac{1}{2}} = 6^{\frac{1}{2}} = \sqrt{6}$$

| power of a product is the product of the powers |

Please note, however that a power of a sum is **not** the sum of the powers. In particular

| $\sqrt{9+16}$ is NOT $\sqrt{9} + \sqrt{16}$ |
| $\downarrow$ $\downarrow$ |
| 5 $3 + 4$ |

SOLUTION: To rationalize the denominator of, say $\frac{5}{\sqrt{2}}$, is to express the fraction without a square root in the denominator. To do so, simply multiply its denominator and numerator by $\sqrt{2}$:

$$\frac{5}{\sqrt{2}} \cdot \frac{\sqrt{2}}{\sqrt{2}} = \frac{5\sqrt{2}}{(\sqrt{2})^2} \underset{\text{see margin}}{\uparrow} \frac{5\sqrt{2}}{2}$$

In addition (see margin):

$$\sqrt{\frac{2}{3}} = \frac{\sqrt{2}}{\sqrt{3}} = \frac{\sqrt{2}}{\sqrt{3}} \cdot \frac{\sqrt{3}}{\sqrt{3}} = \frac{\sqrt{2 \cdot 3}}{(\sqrt{3})^2} = \frac{\sqrt{6}}{3}$$

Leading us to:

$$\frac{5}{\sqrt{2}} + \sqrt{\frac{2}{3}} = \frac{5\sqrt{2}}{2} + \frac{\sqrt{6}}{3} = \frac{15\sqrt{2}}{6} + \frac{2\sqrt{6}}{6} = \frac{15\sqrt{2} + 2\sqrt{6}}{6}$$

Can you now manage Question 6.7:

Simplify: $\dfrac{2}{\sqrt{3}} - \sqrt{\dfrac{2}{5}}$ Answer: $\dfrac{10\sqrt{3} - 3\sqrt{10}}{15}$

If so, go to Question 6.8. If not:

6.7 Solve: *Click-Video*

(a) $\dfrac{3}{\sqrt{2}} + \sqrt{\dfrac{1}{3}}$ (b) $-\sqrt{\dfrac{5}{3}} + \dfrac{2}{\sqrt{5}} - \dfrac{1}{\sqrt{2}}$

If you still can't solve Question 6.7: **Go to the tutoring center.**

Question 6.8

Calculators use the letter "E" (for exponent) in exhibiting scientific notation, as in:

$3.25\text{E}\,8$ (for 3.25×10^8)

The effect of multiplying 4.021 by 10 is to move the decimal point 1 place to the right:

$4.021 \times 10^1 = 40.21$

If you multiply 4.021 by 100, then the decimal point will be moved 2 places to the right:

$4.021 \times 10^2 = 402.1$

and so on.

Express 325,000,000 in scientific notation.

The correct answer is 3.25×10^8. If you got it, move on to Question 6.9. If not, consider the following example:

EXAMPLE 6.8   Express in scientific notation.

(a) 40,210,000   (b) 0.0000023

SOLUTION: Any number can be represented in the form $a \times 10^n$ where $1 \leq a < 10$, n is an integer, and $\times$ denotes multiplication. This is called **scientific notation**, and is particularly useful for expressing very large and very small quantities. In particular (see margin):

(a) $40{,}210{,}000 = 4.021 \times 10^7$

| move decimal point 7 places to the **right** | — 40,210,000

7 places $\longrightarrow$

The effect of dividing 2.3 by 10 is to move the decimal point 1 place to the left:

$$2.3 \times 10^{-1} = 0.23$$

If you divide 2.3 by 100, then the decimal point will be moved 2 places to the left:

$$2.3 \times 10^{-2} = 0.023$$

and so on.

(b) (see margin) $0.0000023 = 2.3 \times 10^{-6}$

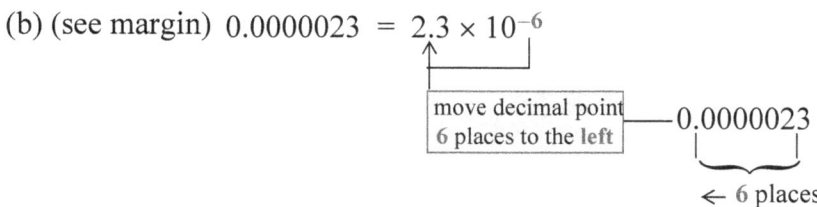

move decimal point 6 places to the **left** — 0.0000023

← 6 places

Can you now manage Question 6.8:

Express 325,000,000 in scientific notation.   Answer: 3.25×10^8

If so, go to Question 6.9. If not:

6.8 Express in scientific notation   *Click-Video*

(a) 23,540,0000   (b) 0.000123

If you still can't solve Question 6.8: **Go to the tutoring center.**

Question 6.9

Express 1.05×10^{-4} in decimal notation.

The correct answer is 0.000105. If you did not get it, consider the following example **here**:

EXAMPLE 6.9   Express in decimal notation.

(a) 3.271×10^{-5} (b) 5.01×10^7

SOLUTION:

(a)   move decimal point **5** places to the **left**

$$3.271 \times 10^{-5} = 0.00003271$$

(b)   move decimal point **7** units to the **right**

$$5.01 \times 10^7 = 50,100,000.0$$

Can you now manage Question 6.9:

Express 1.05×10^{-4} in decimal notation.   Answer: 0.000105

If not:

6.8 Express in decimal notation   *Click-Video*

(a) 2.301×10^{-5} (b) 3.12×10^4

If you still can't solve Question 69: **Go to the tutoring center.**

	SUMMARY
FUNCTIONS	$f(x) = 2x + 5$ $f\boxed{} = 2\boxed{} + 5$ $f(3) = 2 \cdot 3 + 5 = 11$ $f(c) = 2 \cdot c + 5 = 2c + 5$ $f(3t) = 2 \cdot 3t + 5 = 6t + 5$ $f(x^2 + 3) = 2(x^2 + 3) + 5 = 2x^2 + 11$
PERCENTAGE PROBLEMS	What is 22% of 90 $x = \dfrac{22}{100} \cdot 90$ 27 is **what percent** of 18 $27 = \dfrac{x}{100} \cdot 18$ 15 is 75% of what $15 = \dfrac{75}{100} \cdot x$
PRINCIPAL SQUARE ROOT	For $a \geq 0$, the expression $\sqrt{a}$ or $a^{\frac{1}{2}}$ is called the **principal square root of a,** and is that **non-negative** number which when squared equals a.
WARNING	While it is true that for any nonnegative numbers a and b: $\sqrt{ab} = \sqrt{a}\sqrt{b}$ and $\sqrt{\dfrac{a}{b}} = \dfrac{\sqrt{a}}{\sqrt{b}}$ (for $b \neq 0$) in general, the square root of a sum is NOT equal to the sum of the square roots: $\boxed{\sqrt{16 + 9}} = \sqrt{25} = 5$ while $\boxed{\sqrt{16} + \sqrt{9}} = 4 + 3 = 7$ **Not equal**
SCIENTIFIC NOTATION	Any number can be represented in the form $a \times 10^n$ where $1 \leq a < 10$, n is an integer, and $\times$ denotes multiplication. This is called **scientific notation**, and is particularly useful for expressing very large and very small quantities. For example: $0.0000023 = 2.3 \times 10^{-6}$ and $40{,}210{,}000.0 = 4.021 \times 10^7$ move decimal point 6 places to the **left** move decimal point 7 places to the **right**

ADDITIONAL PROBLEMS

1.1 Determine $f(-5)$ if $f(x) = \dfrac{-x^2}{3x+5}$

Answer: $\dfrac{5}{2}$

1.2 Determine $f\left(\dfrac{2}{5}\right)$ if $f(x) = 3x^2 - 2x + \dfrac{1}{3}$

Answer: $\dfrac{1}{75}$

2.1 Determine $f(a+2)$ if $f(x) = x^2 + x - 1$

Answer: $a^2 + 5a + 5$

2.2 Determine $f(x+h)$ if $f(x) = \dfrac{3x}{2x^2 - 1}$

Answer: $\dfrac{3x + 3h}{2x^2 + 4xh + 2h^2 - 1}$

3.1 What is 9% of 80?

Answer: $\dfrac{36}{5} = 7.2$

3.2 What is 115% of 5?

Answer: $\dfrac{23}{4} = 5.75$

4.1 5 is what percent of 20?

Answer: 25

4.2 20 is what percent of 5?

Answer: 400

5.1 20 is 40 percent of what?

Answer: 50

5.2 165 is 110 percent of what?

Answer: 150

6.1 Evaluate: $\dfrac{-\sqrt{4} + \sqrt{9}}{\sqrt{49}}$

Answer: $\dfrac{1}{7}$

6.2 Evaluate: $\dfrac{(64)^{\frac{1}{2}} \cdot 4^{\frac{1}{2}}}{25^{\frac{1}{2}} + 4^{\frac{1}{2}}}$

Answer: $\dfrac{16}{7}$

7.1 Evaluate: $\dfrac{\sqrt{25 \cdot 9}}{\sqrt{\dfrac{25}{9}}}$

Answer: 9

7.2 Evaluate: $\dfrac{\left(\dfrac{25 \cdot 4}{9}\right)^{\frac{1}{2}}}{(16+9)^{\frac{1}{2}}}$

Answer: $\dfrac{2}{3}$

8.1 Express 352,000,000 in scientific notation.

Answer: 3.52×10^8

8.2 Express 0.000205 in scientific notation.

Answer: 2.05×10^{-4}

9.1 Express 1.07×10^4 in decimal notation.

Answer: 10,700

9.2 Express 1.921×10^{-4} in decimal notation.

Answer: 0.0001921

Sample Test 6
SUPPLEMENT

Roughly speaking, a **function** is a rule assigning to each element of one set (collection of objects) exactly one element of another set. There is the age function, for example, which assigns to each individual his or her age; the enrollment function, which associates with a course the number of students enrolled in that course; the grade function, which associates a grade to each student in the course; the profit versus production function; the temperature function; and the list goes on. In short, functions are so important that we simply could not function without them, and the same can be said for their pictorial representations: graphs. Just open any newspaper and you will find graphs which, in one way or another, compactly represent functions: graphs that depict the rise and fall of new housing starts in recent years; graphs representing stock values over a period of time; graphs representing the unemployment rate, population growth, and so on.

Functions can often be described in terms of mathematical expressions. One may, for example, speak of the function:

In the expression:

$$y = f(x)$$

x is said to be the **independent variable**, and y the **dependent variable** (the value of y depends on the value of x).

$$f(x) = 2x + 3$$

(Read: f-of-x equals two-x plus three)

The above function assigns to $x = 5$ the function value:

$$f(5) = 2 \cdot 5 + 3 = 10 + 3 = 13$$

Similarly:

$$f(0) = 2 \cdot 0 + 3 = 3, \text{ and } f(-10) = 2(-10) + 3 = -17$$

By the same token, for the function:

$$g(x) = x^2 + 5x - 1$$

we have:

$$g(2) = 2^2 + 5 \cdot 2 - 1 = 4 + 10 - 1 = 13$$

and: $g(-3) = (-3)^2 + 5(-3) - 1 = 9 - 15 - 1 = -7$

To **evaluate** a function at $x = c$ is to find the value $f(c)$.

EXAMPLE 6.1   Evaluate the function:

$$f(x) = \frac{-x^2}{x + 1}$$

at $x = 5$ and at $x = -2$.

Note: There is a big difference between -5^2 and $(-5)^2$:

$(-5)^2 = (-5)(-5) = 25$
but:
$-5^2 = -(5)(5) = -25$
↑ that "-" is not being squared!

SOLUTION:

$$f(5) = \frac{-5^2}{5+1} = -\frac{25}{6}$$

and:

$$f(-2) = \frac{-(-2)^2}{-2+1} = \frac{-4}{-1} = 4$$

We want to emphasize the fact that the variable x in, say, the function $f(x) = 2x + 5$ is a place holder, a "box" if you will — a box that can hold any meaningful expression:

$$f(x) = 2x + 5$$

$$f\boxed{} = 2\boxed{} + 5$$

In particular:

$$f(3) = 2 \cdot 3 + 5 = 11$$
$$f(c) = 2 \cdot c + 5 = 2c + 5$$
$$f(3t) = 2 \cdot 3t + 5 = 6t + 5$$

and:

$$f(x^2 + 3) = 2(x^2 + 3) + 5 = 2x^2 + 11$$

CHECK YOUR UNDERSTANDING 6.1

For $f(x) = 3x - 5$, determine:

(a) $f(-2)$ 　　(b) $f(t+1)$ 　　(c) $f(-2x+1)$ 　　(d) $f\left(\frac{-2}{x}\right)$

Answers: (a) -11 　(b) $3t - 2$ 　(c) $-6x - 2$ 　(d) $-\frac{6}{x} - 5$

Solution: Page 113.

There are a couple of key words that often translate into mathematical forms. One such word is "of," which often translates into "times." For example:

"one-third of nine" translates to: $\frac{1}{3} \cdot 9$

Another key word is actually a whole class of words, namely any form of the verb "to be." Such words generally translate into "equal." for example:

"one-third of nine is three" translates to: $\frac{1}{3} \cdot 9 = 3$

By definition, x percent, written $x\%$, simply represents the fraction $\frac{x}{100}$; and x percent of something means $\frac{x}{100}$ times that something. For example, 35 percent of 130 is

$$\frac{35}{100}(130) = (.35)130 = 45.5$$

Basically, there are three types of percentage problems:

35% of 130 is what	15% of what is 150	what % of 80 is 60
$\frac{35}{100} \cdot 130 = x$	$\frac{15}{100} \cdot x = 150$	$\frac{x}{100} \cdot 80 = 60$
Answer:	Answer:	Answer:
$x = (.35)130 = 45.5$	$x = \dfrac{150}{\frac{15}{100}}$	$x = \dfrac{(60)(100)}{80}$
	$= 150 \cdot \dfrac{100}{15} = 1000$	$= 75$

CHECK YOUR UNDERSTANDING 6.2

(a) What is $\frac{1}{2}$ of 96?   (b) What is $\frac{1}{2}$ of $\frac{4}{5}$ of 120?

(c) What is 20% of 360?   (d) 18 is what percent of 160?

(e) What percent of 85 is 119?

Answers: (a) 48   (b) 48   (c) 72   (d) $\frac{45}{4}$%   (e) 140%

Solution: Page 113.

There are two numbers that when squared equal 4: 2 and -2:

$$2^2 = 4 \quad \text{and} \quad (-2)^2 = 4$$

SQUARE ROOT AND PRINCIPAL SQUARE ROOT

We say that 2 and -2 are **square roots** of 4. To distinguish between the two roots, we call 2 the **principal square root** of 4 and denote it by the symbol $\sqrt{4}$ or by its exponent form: $4^{\frac{1}{2}}$. Similarly, for any nonnegative number a, the principal square root of a is that nonnegative number, denoted by $\sqrt{a}$ or by $a^{\frac{1}{2}}$, such that:

$$(\sqrt{a})^2 = a \qquad \text{or} \qquad \left(a^{\frac{1}{2}}\right)^2 = a$$

Please note that (in the real number system) you can only take the square root of a nonnegative number, and if a is nonnegative, so is $\sqrt{a}$. For example:

$$\sqrt{16} = 4 \text{ (and } \textbf{not} \pm 4) \qquad \text{while} \quad \sqrt{-16} \text{ is not defined}$$

there is no (real) number which when squared is -16

What is $\sqrt{17}$? It is exactly what it claims to be; namely that number which when squared equals 17. Unlike $\sqrt{16}$ which conveniently turns out to be 4, $\sqrt{17}$ has no "nicer" representation (just like $\frac{16}{2} = 8$, while $\frac{17}{2}$ has no "nicer" representation). A calculator will gladly give you a decimal **approximation** for $\sqrt{17}$, something like: 4.1231056; but 4.1231056 is **not** exactly $\sqrt{17}$, since the square of 4.1231056 is not exactly equal to 17.

WARNING: While it is true that;

$$\sqrt{16 \cdot 9} = \sqrt{16}\sqrt{9} = 4 \cdot 3 \text{ or: } (16 \cdot 9)^{\frac{1}{2}} = 16^{\frac{1}{2}} \cdot 9^{\frac{1}{2}} = 4 \cdot 3$$

The power of a product **IS** the product of the powers.

But

The power of a sum is **NOT** the sum of the powers.

and that: $\sqrt{\dfrac{16}{9}} = \dfrac{\sqrt{16}}{\sqrt{9}} = \dfrac{4}{3}$ or: $\left(\dfrac{16}{9}\right)^{\frac{1}{2}} = \dfrac{16^{\frac{1}{2}}}{9^{\frac{1}{2}}} = \dfrac{4}{3}$

It is **NOT** true that $\sqrt{16 + 9}$ equals $\sqrt{16} + \sqrt{9}$

Since: $\sqrt{16 + 9} = \sqrt{25} = 5$ while $\sqrt{16} + \sqrt{9} = 4 + 3 = 7$

EXAMPLE 6.2   Evaluate:

$$\text{(a) } \left(-16^{\frac{1}{2}} - \sqrt{25}\right)^2 \qquad \text{(b) } \frac{3^{-2} \cdot \left(\sqrt{\frac{7}{9}} - \frac{1}{3}\right)^3}{4^{1/2}}$$

SOLUTION: (a) $\left(-16^{\frac{1}{2}} - \sqrt{25}\right)^2 = (-4 - 5)^2 = (-9)^2 = 81$

$$\text{(b) } \frac{3^{-2} \cdot \left(\sqrt{\frac{7}{9}} - \frac{1}{3}\right)^3}{4^{1/2}} = \frac{\left(\sqrt{\frac{7}{9}} - \frac{3}{9}\right)^3}{3^2 \cdot 2} = \frac{\left(\sqrt{\frac{4}{9}}\right)^3}{9 \cdot 2} = \frac{\left(\frac{2}{3}\right)^3}{18}$$

$$= \frac{\frac{8}{27}}{\frac{18}{1}} = \frac{8}{27} \cdot \frac{1}{18} = \frac{4}{27 \cdot 9} = \frac{4}{243}$$

CHECK YOUR UNDERSTANDING 6.3

Simplify:

(a) $\sqrt{5^2 - \sqrt{9}}$

(b) $\dfrac{64^{\frac{1}{2}}}{\sqrt{2^2 + 32}}$

Answers: (a) $\sqrt{22}$ (b) $\frac{4}{3}$

Solution: Page 113.

To rationalize the denominator of, say $\dfrac{3}{\sqrt{7}}$, is to express the fraction without a square root in the denominator. To do so, simply multiply its denominator and numerator by $\sqrt{7}$:

Recall that for $a \geq 0$, $\sqrt{a}$ is defined to be that number which when squared equals a. In particular $(\sqrt{7})^2 = 7$,

$$\frac{3}{\sqrt{7}} \cdot \frac{\sqrt{7}}{\sqrt{7}} = \frac{3\sqrt{7}}{(\sqrt{7})^2} \underset{\underset{\text{see margin}}{\uparrow}}{=} \frac{3\sqrt{7}}{7}$$

In addition:

$$\sqrt{\frac{3}{7}} = \frac{\sqrt{3}}{\sqrt{7}} = \frac{\sqrt{3}}{\sqrt{7}} \cdot \frac{\sqrt{7}}{\sqrt{7}} = \frac{\sqrt{3 \cdot 7}}{(\sqrt{7})^2} = \frac{\sqrt{21}}{7}$$

EXAMPLE 6.3   Simplify:

$$\frac{2}{\sqrt{3}} + \sqrt{\frac{9}{5}}$$

SOLUTION: $\dfrac{2}{\sqrt{3}} + \sqrt{\dfrac{9}{5}} = \dfrac{2}{\sqrt{3}} + \dfrac{\sqrt{9}}{\sqrt{5}} = \dfrac{2}{\sqrt{3}} + \dfrac{3}{\sqrt{5}}$

$$= \frac{2}{\sqrt{3}} \frac{\sqrt{3}}{\sqrt{3}} + \frac{3}{\sqrt{5}} \frac{\sqrt{5}}{\sqrt{5}}$$

$$= \frac{2\sqrt{3}}{3} + \frac{3\sqrt{5}}{5} = \frac{10\sqrt{3} + 9\sqrt{5}}{15}$$

CHECK YOUR UNDERSTANDING 6.4

Simplify:

(a) $\sqrt{\dfrac{2}{3}} - \dfrac{3}{\sqrt{7}}$

(b) $\dfrac{-3}{\sqrt{2}} + \dfrac{3}{\sqrt{2}}\sqrt{\dfrac{2}{5}} - \dfrac{1}{2\sqrt{3}}$

Answers: (a) $\dfrac{7\sqrt{6} + 9\sqrt{7}}{21}$ (b) $\dfrac{45\sqrt{2} + 18\sqrt{5} - 5\sqrt{3}}{30}$

Solution: Page 113.

SCIENTIFIC NOTATION

Any number can be expressed as the product of a number between 1 and 10 and an integer power of 10. When a number is expressed in such a form, it is said to be represented in **scientific notation,** For example:

$$3865 = 3.865 \times 10^3 \text{ since } 10^3 = 1000 \text{ and } (3.865)(1000) = 3865$$

and:

$$0.0795 = 7.95 \times 10^{-2} \text{ since } 10^{-2} = \frac{1}{100} \text{ and } \frac{7.95}{100} = 0.0795$$

In general:

> Multiplying a number in decimal form by 10^n will result in moving the decimal point n positions to the right if the integer n is positive, or n positions to the left if n is negative.

In scientific notation, the symbol "×" is used to indicate multiplication.

For example:

$$32.356 \times 10^2 = 3235.6$$

$$32.356 \times 10^6 = 32356000$$

$$32.356 \times 10^{-2} = 0.32356$$

$$32.356 \times 10^{-6} = 0.000032356$$

And:

$$1879.32 = 1.87932 \times 10^3$$
$$0.00791 = 7.91 \times 10^{-3}$$
$$35600000 = 3.56 \times 10^7$$
$$0.000000011 = 1.1 \times 10^{-8}$$

CHECK YOUR UNDERSTANDING 6.5

Express in scientific notation:

  (a) 120, 200, 000   (b) 0.00019

Express in decimal notation:

  (c) 3.007×10^5 (d) 3.007×10^{-3}

Answers: (a) 1.202×10^8 (b) 1.9×10^{-4}
              (c) 300,700   (d) 0.003007
Solution: Page 114.

APPENDIX A
CHECK YOUR UNDERSTANDING SOLUTIONS

1. Rational Numbers

__CYU 1.1__   (a) $\dfrac{24}{30} = \dfrac{2^3 \cdot 3}{2 \cdot 3 \cdot 5} = \dfrac{2^2}{5} = \dfrac{4}{5}$

(b) $\dfrac{-92}{44} = -\dfrac{2^2 \cdot 23}{2^2 \cdot 11} = \dfrac{-23}{11} \ \left(\text{or: } -\dfrac{23}{11}\right)$

(c) $\dfrac{15(2+c)}{10(c+2)} = \dfrac{3 \cdot 5(2+c)}{2 \cdot 5(2+c)} = \dfrac{3}{2}$

(d) $\dfrac{7b(a+c)}{b(-a-c)} = \dfrac{7(a+c)}{-(a+c)} = \dfrac{7}{-1} = -7$

__CYU 1.2__   (a) $\left(\dfrac{-5}{21}\right)\left(\dfrac{14}{10}\right) = \dfrac{-5 \cdot 2 \cdot 7}{3 \cdot 7 \cdot 2 \cdot 5} = \dfrac{-1}{3} \ \left(\text{or } -\dfrac{1}{3}\right)$

(b) $\dfrac{9}{5} \cdot \dfrac{15}{8} \cdot 12 \cdot \dfrac{2}{6} = \dfrac{3^2 \cdot 3 \cdot 5 \cdot 2^2 \cdot 3 \cdot 2}{5 \cdot 2^3 \cdot 2 \cdot 3} = \dfrac{2^3 \cdot 3^4 \cdot 5}{2^4 \cdot 3 \cdot 5} = \dfrac{27}{2}$

__CYU 1.3__   (a) $\dfrac{\frac{2}{3}}{\frac{8}{9}} = \dfrac{2}{3} \cdot \dfrac{9}{8} = \dfrac{3}{4}$

(b) $\dfrac{\frac{2}{3}}{9} = \dfrac{\frac{2}{3}}{\frac{9}{1}} = \dfrac{2}{3} \cdot \dfrac{1}{9} = \dfrac{2}{27}$

(c) $\dfrac{2}{\frac{8}{9}} = \dfrac{\frac{2}{1}}{\frac{8}{9}} = \dfrac{2}{1} \cdot \dfrac{9}{8} = \dfrac{9}{4}$

(d) $\dfrac{\frac{2a+2b}{4}}{\frac{a+b}{8}} = \dfrac{2(a+b)}{4} \cdot \dfrac{8}{a+b} = 4$

__CYU 1.4__   (a) $\dfrac{-3}{9} + \dfrac{2}{9} + \dfrac{19}{9} = \dfrac{-3+2+19}{9} = \dfrac{18}{9} = 2$

(b) $\dfrac{3}{a} + \dfrac{2}{a} + \dfrac{-5}{a} = \dfrac{3+2-5}{a} = \dfrac{0}{a} = 0$ (assuming $a \ne 0$)

(c) $\dfrac{-3a}{4} + \dfrac{2(a-1)}{4} + \dfrac{5}{4} = \dfrac{-3a+2a-2+5}{4} = \dfrac{-a+3}{4}$

__CYU 1.5__   (a) $\dfrac{-3}{18} + \dfrac{1}{12} + \dfrac{3}{8} = \dfrac{-3}{2 \cdot 3^2} + \dfrac{1}{2^2 \cdot 3} + \dfrac{3}{2^3} = \dfrac{-1}{2 \cdot 3} + \dfrac{1}{2^2 \cdot 3} + \dfrac{3}{2^3}$

building up to the LCD: $2^3 \cdot 3$: $= \dfrac{-1(2^2)}{(2 \cdot 3)(2^2)} + \dfrac{1(2)}{(2^2 \cdot 3)(2)} + \dfrac{3(3)}{(2^3)(3)}$

$= \dfrac{-4}{2^3 \cdot 3} + \dfrac{2}{2^3 \cdot 3} + \dfrac{9}{2^3 \cdot 3} = \dfrac{7}{24}$

(b) $\dfrac{-2}{15} + \dfrac{1}{5} + \dfrac{1}{3} = \dfrac{-2}{15} + \dfrac{1(3)}{15} + \dfrac{1(5)}{15} = \dfrac{-2+3+5}{15} = \dfrac{6}{15} = \dfrac{2}{5}$

CYU 1.6 (a) $\dfrac{\left(-2+\frac{1}{3}\right)\left(1-\frac{1}{2}\right)}{1+\frac{1}{2}} = \dfrac{\left(-\frac{6}{3}+\frac{1}{3}\right)\left(\frac{2}{2}-\frac{1}{2}\right)}{\frac{2}{2}+\frac{1}{2}} = \dfrac{\left(\frac{-5}{3}\right)\left(\frac{1}{2}\right)}{\frac{3}{2}} = \left(\frac{-5}{3}\right)\left(\frac{1}{2}\right)\left(\frac{2}{3}\right) = \dfrac{-5}{9}$

(b) $\dfrac{\frac{3b}{a+b} - \frac{a}{2a+2b}}{\frac{4}{a+b}} = \left[\dfrac{6b}{2(a+b)} - \dfrac{a}{2(a+b)}\right]\dfrac{a+b}{4} = \dfrac{6b-a}{2(a+b)}\dfrac{(a+b)}{4} = \dfrac{6b-a}{8}$

2. Integer Exponents

CYU 2.1 (a) $(1+4)^2 + 3^2 - (2\cdot 3)^2 + 2^{2+3} = 5^2 + 3^2 - 6^2 + 2^5 = 25 + 9 - 36 + 32 = 30$

(b) $\left(\frac{1}{2}-1\right)^2 + 8\cdot 2^2 - (2-3)^2 - [-2^2]^3 = \left(-\frac{1}{2}\right)^2 + 8\cdot 4 - 1 - (-4)^3$

$$= \frac{1}{4} + 32 - 1 + 64 = \frac{1}{4} + 95$$

$$= \frac{1}{4} + \frac{95(4)}{4} = \frac{1+380}{4} = \frac{381}{4}$$

(c) $\left(2+\frac{1}{2}\right)^2 - \left[-1-\frac{1}{3}\right]^2 + \frac{1}{9} = \left(\frac{5}{2}\right)^2 - \left(-\frac{4}{3}\right)^2 + \frac{1}{9} = \frac{25}{4} - \frac{16}{9} + \frac{1}{9}$

$$= \frac{25}{4} - \frac{15}{9} = \frac{25}{4} - \frac{5}{3}$$

$$= \frac{25\cdot 3}{4\cdot 3} - \frac{5\cdot 4}{3\cdot 4}$$

$$= \frac{75-20}{12} = \frac{55}{12}$$

(d) $(3a)^2 - 2a^2 + 2^2a + a(2^2+a) = 9a^2 - 2a^2 + 4a + 4a + a^2 = 8a^2 + 8a$

CYU 2.2 (a) $\dfrac{2^3\left(\frac{1}{2}\right)^2}{\left(1+\frac{1}{2}\right)^{-1}} = \dfrac{\frac{2^3}{2^2}}{\frac{1}{\frac{3}{2}}} = \dfrac{\frac{2}{1}}{\frac{1}{\frac{3}{2}}} = \dfrac{2}{1}\cdot\dfrac{\frac{3}{2}}{1} = 2\cdot\dfrac{3}{2} = 3$

(b) $\left[2^{-1}+\left(\frac{1}{2}\right)^{-2}\right]^{-1} = \left(\dfrac{1}{2}+\dfrac{1}{\left(\frac{1}{2}\right)^2}\right)^{-1} = \left(\dfrac{1}{2}+\dfrac{1}{\frac{1}{4}}\right)^{-1} = \left(\dfrac{1}{2}+4\right)^{-1} = \left(\dfrac{9}{2}\right)^{-1} = \dfrac{1}{\frac{9}{2}} = \dfrac{2}{9}$

(c) $\left[\dfrac{5^{-1}+\left(3-\dfrac{1}{2}\right)^{-2}}{5\left(\dfrac{1}{3}-\dfrac{5}{9}\right)^{17}}\right]^{0}+\dfrac{1}{2} = 1+\dfrac{1}{2} = \dfrac{3}{2}$

(d) $\dfrac{(2a)^{-2}(-b)^2}{a(b)^{-1}} = \dfrac{\dfrac{1}{(2a)^2}b^2}{a\cdot\dfrac{1}{b}} = \dfrac{\dfrac{b^2}{4a^2}}{\dfrac{a}{b}} = \dfrac{b^2}{4a^2}\cdot\dfrac{b}{a} = \dfrac{b^3}{4a^3}$ (providing $a\ne 0$ and $b\ne 0$)

3. Algebraic Expressions

__CYU 3.1__   (a) $\dfrac{(-2a)^2}{5b}\cdot\dfrac{-b^2}{10}\cdot\dfrac{50}{4b} = \dfrac{4a^2(-b^2)(50)}{5b(10)(4b)} = \dfrac{4(50)a^2(-b^2)}{4(50)b^2} = -a^2$

(b) $\dfrac{(-abc)^2}{(ab)^2+abc} = \dfrac{a^2b^2c^2}{ab(ab+c)} = \dfrac{abc^2}{ab+c}$

(c) $\dfrac{\dfrac{2a+2b}{4}}{\dfrac{a+b}{8}} = \dfrac{2(a+b)}{4}\cdot\dfrac{8}{a+b} = 4$

__CYU 3.2__   (a) $\dfrac{3}{c}+\dfrac{-6}{bc}+\dfrac{a}{2b} = \dfrac{3(2b)}{2bc}+\dfrac{-6(2)}{2bc}+\dfrac{a(c)}{2bc} = \dfrac{6b-12+ac}{2bc}$

(b) $\dfrac{3}{2x+6}-\dfrac{x}{x+3}+\dfrac{1}{4} = \dfrac{3}{2(x+3)}-\dfrac{x}{x+3}+\dfrac{1}{4} = \dfrac{3(2)}{4(x+3)}-\dfrac{x(4)}{4(x+3)}+\dfrac{1(x+3)}{4(x+3)}$

$\qquad = \dfrac{6-4x+x+3}{4(x+3)} = \dfrac{-3x+9}{4(x+3)} = \dfrac{-3x+9}{4x+12}$

(c) $\dfrac{x}{(-x+2)^2}+\dfrac{3}{x-2} = \dfrac{x}{[-(x-2)]^2}+\dfrac{3}{x-2} = \dfrac{x}{(x-2)^2}+\dfrac{3}{x-2}$

$\qquad = \dfrac{x}{(x-2)^2}+\dfrac{3(x-2)}{(x-2)^2} = \dfrac{x+3x-6}{(x-2)^2} = \dfrac{4x-6}{(x-2)^2}$

__CYU 3.3__   (a) $\dfrac{6x^2(5x+2)^4-3x^3(5x+2)^3}{6(-5x-2)^6} = \dfrac{(5x+2)^3[6x^2(5x+2)-3x^3]}{6[-(5x+2)]^6}$

$\qquad = \dfrac{(5x+2)^3\,3x^2[2(5x+2)-x]}{6(5x+2)^6}$

$\qquad = \dfrac{(5x+2)^3x^2(9x+4)}{2(5x+2)^6}$

$\qquad = \dfrac{x^2(9x+4)}{2(5x+2)^3} = \dfrac{9x^3+4x^2}{2(5x+2)^3}$

(b) $2x(3x-1)^{-1} + 4x^2(3x-1)^{-3} = \dfrac{2x}{3x-1} + \dfrac{4x^2}{(3x-1)^3} = \dfrac{2x(3x-1)^2 + 4x^2}{(3x-1)^3}$

$$= \dfrac{2x(9x^2 - 6x + 1) + 4x^2}{(3x-1)^3}$$

$$= \dfrac{18x^3 - 8x^2 + 2x}{(3x-1)^3}$$

4. Equations and Inequalities

CYU 4.1 (a) $3 - 2x + 5 - x = -4x - 2 + 1$

$-2x - x + 4x = -2 + 1 - 3 - 5$

$x = -9$

(b) $\dfrac{-3x}{5} - \dfrac{x}{2} + 1 = \dfrac{2x+1}{10}$

$-6x - 5x + 10 = 2x + 1$

$-13x = -9$

$x = \dfrac{9}{13}$

CYU 4.2 (a) $7x - 5x - 1 = 2x + 3$

$7x - 5x - 2x = 3 + 1$

$0 = 4$

(contradiction)

(b) $-3x + 7 - x = 2x - 4$

$-6x = -11$

$x = \dfrac{11}{6}$

(conditional)

(c) $\dfrac{x-4}{2} = \dfrac{1}{4}(4x - 12) - \dfrac{x}{2} + 1$

$2(x - 4) = (4x - 12) - 2x + 4$

$2x - 8 = 2x - 8$

(identity)

CYU 4.3 $3x + 2(x - y) = y + 4x + 1$

$3x + 2x - 2y = y + 4x + 1$

$3x + 2x - 4x = y + 2y + 1$

$\boxed{x = 3y + 1} \Rightarrow 3y = x - 1 \Rightarrow \boxed{y = \dfrac{x-1}{3}}$

CYU 4.4 $\dfrac{2x+5}{-2} \le (-3x + 1)$

multiplying by -2

$2x + 5 \ge (-2)(-3x + 1)$

$2x + 5 \ge 6x - 2$

$-4x \ge -7$

dividing by -4

$x \le \dfrac{7}{4}$

CYU 4.5   (a) $x^2 + x - 6 = 0$ (b) $x^2 - 9 = 0$

$\qquad(x+3)(x-2) = 0$ $\qquad(x+3)(x-3) = 0$

$\qquad\quad x = -3, x = 2$ $\qquad\quad x = -3, x = 3$

(c) $x^2 + 5x - 1 = -x^2 + 4x + 2$

$\qquad 2x^2 + x - 3 = 0$

$\qquad (2x+3)(x-1) = 0$

$\qquad\quad x = -\dfrac{3}{2}, x = 1$

CYU 4.6 $x^3 + 3x^2 - 4x - 12 = 0$

$\qquad x^2(x+3) - 4(x+3) = 0$

$\qquad\quad (x^2 - 4)(x+3) = 0$

$\qquad (x+2)(x-2)(x+3) = 0$

$\qquad x = -3, x = -2, x = 2$

5. Lines and Linear Equations

CYU 5.1   (a) $m = \dfrac{2-(-6)}{3-(-1)} = \dfrac{8}{4} = 2$

(b) Moving 1 unit to the right brings you up 2 units. So moving 8 units to the right of $(-1, -6)$, brings. you to: $(-1+8, -6+2\cdot 8) = (7, 10)$.

CYU 5.2   Slope: $m = \dfrac{5-(-6)}{-3-4} = -\dfrac{11}{7}$. It follows that the line is of the form $y = -\dfrac{11}{7}x + b$.

We select the point $(-3, 5)$ to find "b" (when $x = -3$, y must be 5):

$$5 = -\frac{11}{7}(-3) + b \Rightarrow b = 5 - \frac{33}{7} = \frac{2}{7}$$

$$\text{Conclusion: } y = -\frac{11}{7}x + \frac{2}{7}$$

CYU 5.3   (a) Substituting -1 for x and 3 for y in the equation $2x - 4y = 10$ we find that equality does not hold: $2(-1) - 4(3) = -14$ (and not 10). That is good enough to tell us

that $(x = -1, y = 3)$ is not a solution of $\left.\begin{array}{r} 2x - 4y = 10 \\ -x + 5y = -8 \end{array}\right\}$.

(b) Substituting 3 for x and -1 for y in the equations $2x - 4y = 10$, $-x + 5y = -8$ we find that both are satisfied:

$$2(3) - 4(-1) = 10 \text{ and } -3 + 5(-1) = -8$$

CYU 5.4   (a) $2x - 7y = 18 \longrightarrow 2(-5y - 8) - 7y = 18$

$x + 5y = -8 \Rightarrow x = -5y - 8 \qquad\qquad -17y = 34$

$$\boxed{y = -2}$$

From $x = -5y - 8$: $x = -5(-2) - 8 \Rightarrow \boxed{x = 2}$

(b) $\left.\begin{array}{l} \dfrac{y+x}{3} + 3y = \dfrac{1}{6} \\[2mm] \dfrac{2x}{5} - \dfrac{y+1}{10} = \dfrac{1}{10} \end{array}\right\} \Rightarrow \left.\begin{array}{l} 2(y+x) + 18y = 1 \\[2mm] 4x - (y+1) = 1 \end{array}\right\} \Rightarrow$ $2x + 20y = 1$ (*)

$4x - y = 2 \Rightarrow y = 4x - 2$ (**)

substituting in (*)

$2x + 20(4x - 2) = 1$

$82x = 41$

$$\boxed{x = \dfrac{1}{2}}$$

from (**): $y = 4\left(\dfrac{1}{2}\right) - 2 \Rightarrow \boxed{y = 0}$

CYU 5.5   (a) $4x + 5y = -17 \longrightarrow 4x + 5(-1) = -17$

$\dfrac{4x - 2y = -10}{\text{−:} \quad 7y = -7}$ $\qquad\qquad 4x = -12$

$$\boxed{y = -1} \qquad\qquad \boxed{x = -3}$$

(b) $\left.\begin{array}{l} 2x + 3y = \dfrac{3}{2} - x - 2y \\[2mm] -6x - 9y = -3 \end{array}\right\} \Rightarrow \left.\begin{array}{l} 4x + 6y = 3 - 2x - 4y \\[2mm] -6x - 9y = -3 \end{array}\right\} \Rightarrow \left.\begin{array}{l} 6x + 10y = 3 \\[2mm] -6x - 9y = -3 \end{array}\right\}$

$\Rightarrow \dfrac{6x + 10y = 3 \text{ (*)}}{\quad -6x - 9y = -3}$

$\text{+:} \qquad\qquad \boxed{y = 0}$

From (*): $6x + 10(0) = 3$

$$\boxed{x = \dfrac{1}{2}}$$

6. Odds and Ends

CYU 6.1 For $f(x) = 3x - 5$: (a) $f(-2) = 3(-2) - 5 = -6 - 5 = -11$

(b) $f(t + 1) = 3(t + 1) - 5 = 3t + 3 - 5 = 3t - 2$

(c) $f(-2x + 1) = 3(-2x + 1) - 5 = -6x + 3 - 5 = -6x - 2$

(d) $f\left(\dfrac{-2}{x}\right) = 3\left(\dfrac{-2}{x}\right) - 5 = -\dfrac{6}{x} - 5$

CYU 6.2 (a) What is $\dfrac{1}{2}$ of 96? Translation: $x = \dfrac{1}{2} \cdot 96$. Answer: $x = 48$.

(b) What is $\dfrac{1}{2}$ of $\dfrac{4}{5}$ of 120? Translation: $x = \dfrac{1}{2} \cdot \dfrac{4}{5} \cdot 120$. Answer: $x = 48$.

(c) What is 20% of 360? Translation: $x = \dfrac{20}{100} \cdot 360$. Answer: $x = 72$.

(d) 18 is what percent of 160? Translation: $18 = \dfrac{x}{100} \cdot 160$. Answer: $x = \dfrac{45}{4}\%$.

(e) What percent of 85 is 119? Translation: $\dfrac{x}{100} \cdot 85 = 119$. Answer: $x = 140\%$.

CYU 6.3 (a) $\sqrt{5^2 - \sqrt{9}} = \sqrt{25 - 3} = \sqrt{22}$

(b) $\dfrac{64^{\frac{1}{2}}}{\sqrt{2^2 + 32}} = \dfrac{8}{\sqrt{4 + 32}} = \dfrac{8}{\sqrt{36}} = \dfrac{8}{6} = \dfrac{4}{3}$

CYU 6.4 (a) $\sqrt{\dfrac{2}{3}} - \dfrac{3}{\sqrt{7}} = \dfrac{\sqrt{2}}{\sqrt{3}}\dfrac{\sqrt{3}}{\sqrt{3}} - \dfrac{3}{\sqrt{7}}\dfrac{\sqrt{7}}{\sqrt{7}} = \dfrac{\sqrt{6}}{3} + \dfrac{3\sqrt{7}}{7} = \dfrac{7\sqrt{6} + 9\sqrt{7}}{21}$

(b) $\dfrac{-3}{\sqrt{2}} + \dfrac{3}{\sqrt{2}}\sqrt{\dfrac{2}{5}} - \dfrac{1}{2\sqrt{3}} = \dfrac{-3}{\sqrt{2}} + \dfrac{3}{\sqrt{2}}\dfrac{\sqrt{2}}{\sqrt{5}} - \dfrac{1}{2\sqrt{3}}$

$= \dfrac{-3}{\sqrt{2}}\dfrac{\sqrt{2}}{\sqrt{2}} + \dfrac{3}{\sqrt{5}}\dfrac{\sqrt{5}}{\sqrt{5}} - \dfrac{1}{2\sqrt{3}}\dfrac{\sqrt{3}}{\sqrt{3}} = \dfrac{-3\sqrt{2}}{2} + \dfrac{3\sqrt{5}}{5} - \dfrac{\sqrt{3}}{6}$

$= \dfrac{-3\sqrt{2}(15)}{30} + \dfrac{3\sqrt{5}(6)}{30} - \dfrac{\sqrt{3}(5)}{30}$

$= \dfrac{-45\sqrt{2} + 18\sqrt{5} - 5\sqrt{3}}{30}$

CYU 6.5 (a) $120,200,000 = 1.202 \times 10^8$

move decimal point 8 places to the **right**

(b) $0.00019 = 1.9 \times 10^{-4}$

move decimal point 4 places to the **left**

(c) $3.007 \times 10^5 = 300,700$

move decimal point 5 places to the **right**

(d) $3.007 \times 10^{-3} = 0.003007$

move decimal point 3 places to the **left**